Understanding the Building Regulations

6th edition

Simon Polley

 Routledge
Taylor & Francis Group

LONDON AND NEW YORK

First published 1995
by E & FN Spon

This edition published 2015
by Routledge
2 Park Square, Milton Park, Abingdon, Oxon OX14 4RN

and by Routledge
711 Third Avenue, New York, NY 10017

Routledge is an imprint of the Taylor & Francis Group, an informa business

British Library Cataloguing-in-Publication Data
A catalogue record for this book is available from the British Library

Library of Congress Cataloging-in-Publication Data
Polley, Simon, author.
 Understanding the building regulations/Simon Polley. — Sixth edition.
 pages cm
 Includes bibliographical references and index.
 1. Building laws—England. I. Title.
 KD1140.P65 2014
 343.4207′869—dc23
 2014010976

ISBN: 978-0-415-71742-7 (pbk)
ISBN: 978-1-315-75512-0 (ebk)

Typeset in Sabon and Gill Sans
by Florence Production Ltd, Stoodleigh, Devon, UK

MIX
Paper from
responsible sources
FSC® C013056
www.fsc.org

Printed and bound in Great Britain by
TJ International Ltd, Padstow, Cornwall

Contents

Preface

The purpose of this book is to introduce the reader to the current system of building control in England and Wales, based on the Building Regulations 2010 and all the supporting Approved Documents.

The book starts with a brief history lesson in building control and how it has developed over the years. Chapters are then devoted to the Building Regulations and each of the Approved Documents. The application of Building Regulation Requirements and the guidance contained in Approved Documents are discussed and illustrated in a straightforward and logical manner to enable the text to be utilized as a reference source for members of the design and construction team and those who require a knowledge of building control. For information and reference, the actual wording of each Building Regulation Requirement has been stated at the beginning of each chapter.

Since the book represents a simplification of the Building Regulations and Approved Documents it should not be regarded as a replacement but as a standalone support text to the original documents. For further detailed advice and guidance the services of an approved inspector, Local Authority building control department or a building regulation consultancy should be sought.

Simon Polley FBEng MRICS MIFireE
Managing Director, BRCS (Building Control) Limited
Synergy Centre, 5 Hoffmanns Way
Chelmsford, Essex CM1 1GU

Acknowledgements

The author would like to thank the following people for their assistance in producing this book:

Karl Webb-Thomas, Regional Manager, BRCS (Building Control) Limited.

Cartoons and illustrations by 'Bill' Brignell, Cartoonist and Illustrator.

Drawings and diagrams by Barry J. Powell, Architectural Technician.

Introduction

In London in 1189, Henry Fitz Ailwyn was appointed Mayor. His London Assize was virtually the first London Building Act – the first Building Regulations. As an example, if neighbours agreed to build a party wall between their adjoining properties it had to be 3 ft thick and 16 ft high. Unfortunately, no powers existed to enforce this requirement or impose penalties.

Archaeological ruins and debris show that, prior to this date, failures occurred, and that building codes were used mainly as deterrents. In 2000 BC, for example, in Egypt, the ultimate deterrent was introduced as a regulation: if a man died because of a building failure then the builder himself would be held liable and put to death.

In Roman times, rules and regulations were drawn up and enforcement was introduced to try and reduce the possibility of failure, which has been the main theme of regulations and enforcement to date.

Returning to this country, the seventeenth century saw the first Acts to cover the whole of England, although it is the year 1666 that will be remembered, especially by Charles II. In the early hours of Sunday, 1 September the Great Fire of London started at the King's baker in Pudding Lane: four-fifths of the City was destroyed.

In the aftermath, the City appointed four surveyors to draft Regulations, which were duly embodied in the Rebuilding Act of 1667. They included:

- the utilization of four different 'purpose groups';
- minimum storey heights;
- party/external wall thicknesses;
- space separation;
- rainwater pipe provisions.

These regulations for the first time were well detailed, and City Viewers – the first building control surveyors – were appointed to enforce compliance. The late nineteenth century saw the emergence, as a consequence of the Public Health Act 1875, of the two distinct systems of building control, with the District Surveyor responsible for the London County Council area, and local authorities (under the Public Health Act) for the remainder of the country. The model by-laws used were rather basic, reflecting the comparatively simple construction techniques, although importantly they

did require that suitable plans be submitted for Local Authority approval.

In 1881 the average number of persons per dwelling was eight, and despite the many regulations, domestic building standards both in London and across the country were poor. Not until after the First World War did things start to improve. The complexity of building had started to increase with the wide use of structural steelwork, in lieu of cast iron, and the introduction of reinforced concrete. This brought about a more mathematical analysis of building design, a fact that all later legislation took into account.

The principal Building Acts of the time were the London Building Act 1930, amended in 1935 and 1939, and for the remainder of England and Wales the Public Health Act 1936, from which a single series of model by-laws was issued.

After the Second World War was over, the availability of a large range of codes of practice and British Standard specifications mirrored the ever-changing growth of building technology. The Building By-laws of 1953 first gave the option of 'deemed to satisfy' requirements, and on 1 February

1966 the first national Building Regulations came into operation. These were made under the Public Health Act 1961 and included provisions for:

- structural fire precautions;
- requirements for division or compartment walls;
- fire protection to structural elements;
- sound insulation to walls and floors of dwellings;
- minimum stairway dimensions;
- 'Zones of open space';
- a table of exempted buildings.

The Building Regulations 1972 (basically a metric reissue) and the Building Regulations 1976 were to follow. In April 1980 fees for the submission of plans were introduced for the first time under the Building (Prescribed Fees) Regulations.

In February 1981 a Command Paper – *The future of building control in England and Wales* – was published. It contained the Government's proposals for major changes to building control, and saw the light of day in the form of the Building Act 1984. This consolidated nearly all the previous legislation covering building control and included new proposals for optional privatization, streamlining of the system and redrafting of the Building Regulations. The private option for building control first appeared under the Building (Approved Inspector) Regulations 1985.

The Building Regulations 1985 reflected the contents of the 1976 regulations, but their form and arrangement were drastically altered. The Schedule 1 Requirements were written in general terms requiring reasonable standards of health and safety for persons in or about the building. Reference was made to supporting Approved Documents which, with the exception of B1: Means of escape, were not legally enforceable, as various ways could exist to show compliance.

Following a review of the 1985 regulations, by the Department of the Environment, the Building Regulations 1991 came into force on 1 June 1992, with more than half of the Approved Documents being revised. The Building Regulations (Amendment) Regulations 1994 brought about changes to specific regulations and requirements, which included new Approved Documents for Parts F: Ventilation and L: Conservation of fuel and power, operable from 1 July 1995.

Since 1995 the Department of the Environment, Transport and Regions has issued a number of amendment regulations. The Building Regulations (Amendment) (No. 2) Regulations 1999 resulted in the 2000 edition of Approved Document B. Approved Documents K and N and the approved document to support Regulation 7 were revised, and the 1999 edition of Approved Document M applied to dwellings for the first time.

Two non-departmental approved documents appeared, *Timber intermediate floors for dwellings* (TRADA 1992) and *Basements for dwellings*

(British Cement Association 1997). With the introduction of the Building (Local Authority Charges) Regulations 1998, Local Authority building controls could now determine its own fee scales for building regulation submissions. To act as a guide for the submission of building regulation applications the *Manual to the Building Regulations* was also re-introduced.

During 1996, the author was instrumental in bringing together the Association of Building Engineers and insurance brokers Griffiths and Armour so as to put in place a government-approved insurance scheme for Approved Inspectors. BRCS (Building Control) Limited became one of the first Corporate Approved Inspectors in January 1997 to be able to operate in the commercial field. This paved the way for the private sector alternative to Local Authority building control beyond housing. The Building (Approved Inspectors, etc.) Regulations 2000 consolidated previous regulations and incorporated minor amendments and additional requirements. The building regulations themselves were also consolidated and the Building Regulations 2000 came into force on 1 January 2001.

The Building (Amendment) Regulations 2001 saw the publication of new Parts H, J and L. Requirements for building over public sewers and a revised consultation process with the sewerage undertakers was introduced. The requirements of Part J were extended and Part L now incorporates new requirements for lighting systems, solar overheating and mechanical ventilation. Additional amendment regulations in 2002 gave rise to an extended list of exemptions from submitting applications and the introduction of a new Part E. Sound insulation testing would now be required in certain cases and the scope of Part E extended to include schools and 'rooms for residential purposes'.

On 1 May 2004 Part M became 'Access to and use of buildings'. The needs of disabled people are inherently included in the design guidance which draws from BS: 8300. Also in 2004 revisions were made to Parts A and C and a new Approved Document P: Electrical safety, became effective from 1 January 2005. The next major revisions relate to new approved documents for Parts F and L, and amendments to Part B, splitting it into two separate approved documents. Changes also occurred to the Building Regulations themselves that came into force at the beginning of 2008.

In 2010 the updating of the structural Eurocodes impacted on Approved Documents A and C, revisions were made to Parts F and L, and new Approved Documents were introduced for Parts G and J. The most significant improvement was however the consolidation of the Building Regulations 2010 and the Building (Approved Inspectors etc.) Regulations 2010.

The Building (Amendment) Regulations 2012 and 2013 introduced a number of key alterations and additions to the regulations themselves, mainly relating to energy requirements and alterations to the local authority notification procedure. A new Approved Document K now incorporates overlapping guidance from Approved Document M, which has been updated, and the complete contents of Approved Document N, which is now withdrawn. New Approved Documents L1A and L2A, covering new build projects, have been issued with relatively minor updates made to L1B and L2B. Minor alterations have taken place to the other Approved Documents, with a new Approved Document P reducing coverage for electrical work on dwellings.

Since December 2011 (under the Welsh Ministers (Transfer of Functions) (No. 2) Order 2009), Wales has become responsible for making their own Building Regulations. The opportunity has been taken to clarify the impact of their new Regulations and Approved Document guidance, mainly relating to energy use and the introduction of residential sprinklers. The Approved Documents and associated legislation, for both England and Wales, are always under review; updates can be found at www.gov.uk and www.wales.gov.uk.

Chapter 1

The Building Regulations 2010

The Building Regulations 2010 are a Statutory Instrument, 2010/2214, made under specific sections of the Building Act 1984. They impose requirements on people carrying out certain building work in England and Wales (**excepted energy buildings** only). It is important to note that compliance must be shown with these regulations and not necessarily with the contents of the Approved Documents, which are purely to give 'practical guidance with respect to the requirements of any provision of building regulations' (Section 6 of the Building Act 1984), and their use is not therefore mandatory.

The actual Regulations, which are discussed below, are split into ten parts concluding with six schedules:

- Schedule 1 – the technical requirements expressed in functional terms with which building work must comply;
- Schedule 2 – exempt buildings and work as referred to in Regulation 9;
- Schedule 3 – self-certification schemes and exemptions from requirements to give building notice or deposit full plans;
- Schedule 4 – descriptions of work where no building notice or deposit full plans required;
- Schedule 5 – revocation of regulations;
- Schedule 6 – consequential amendments.

Regulation 1: Citation and commencement

Regulation 2: Interpretation

A number of terms used within the regulations are explained. The significant ones are reiterated for information:

Building – any permanent or temporary building (or part of a building) but not any other kind of structure or erection.

Change to a building's energy status – any change that results in a building becoming a building to which the energy efficiency requirements of these Regulations apply, where previously it was not.

Controlled service or fitting – a service or fitting where Part G, H, J, L or P imposes a requirement.

Dwelling – includes a dwelling-house and a flat.

Dwelling-house – does not include a flat or a building containing a flat.

Energy efficiency requirements – the requirements of Regulations 23, 25A, 25B, 26, 26A [26B and 25C Wales only], 28, 29, 40 and 43, and Part L of Schedule 1.

European technical approval – a favourable technical assessment of the fitness for use of a construction product for the purposes of the Construction Products Directive, issued by a European Technical Approval issuing body.

Excepted energy building is defined in Schedule 2 of the Welsh Ministers (Transfer of Functions) (No.2) Order 2009. In essence this only includes a generating station, and associated electric lines, pipe-lines and storage that are not used as a: residence; shop; office; showroom; canteen; or ancillary outbuildings.

Flat – separate and self-contained premises, including a maisonette, constructed or adapted for residential use and forming part of a building divided horizontally from some other part.

Floor area – the aggregate area of every floor in a building or extension, calculated by reference to the finished internal faces of the walls enclosing the area or, if at any point there is no such wall, by reference to the outermost edge of the floor.

Height – the height of the building measured from the mean adjoining outside ground level to a level half the vertical height of the roof or to the top of any walls or parapets, whichever is higher.

Institution – an institution (whether described as a hospital, home, school or other similar establishment) that is used as living accommodation for, or for the treatment, care or maintenance of, persons:

(a) suffering from disabilities due to illness or old age or other physical or mental incapacity; or

(b) under the age of five years,

where such persons sleep on the premises.

Public building – consisting of or containing:

(a) a theatre, public library, hall or other place of public resort;

(b) a school or other educational establishment [not exempted under Section 4(1)(a) of the Building Act 1984]; or

(c) a place of public worship,

but excluding a shop, storehouse or warehouse, or a dwelling to which members of the public are occasionally admitted.

Renovation – the provision of a new layer in the thermal element, unless solely as a flat roof repair, or the replacement of an existing layer, but excludes decorative finishes.

Room for residential purposes – a room, or a suite of rooms, that is not a dwelling-house or a flat and that is used by one or more persons to live and sleep in; includes a room in a hostel, a hotel, a boarding house, a hall of residence or a residential home, whether or not the room is separated from or arranged in a cluster group with other rooms, but does not include a room in a hospital, or other similar establishment, used for patient accommodation, and, for the purposes of this definition, a 'cluster' is a group of rooms for residential purposes that is:

(a) separated from the rest of the building in which it is situated by a door that is designed to be locked; and

(b) not designed to be occupied by a single household.

In these Regulations, **thermal element** means a wall, floor or roof (but does not include windows, doors, roof windows or roof-lights) that separates a thermally conditioned part of the building (**the conditioned space**) from:

(a) the external environment (including the ground); or

(b) in the case of floors and walls, another part of the building that is:

    (i) unconditioned;

    (ii) an extension falling within Class VII of Schedule 2; or

    (iii) where this paragraph applies, conditioned to a different temperature, and includes all parts of the element between the surface bounding the conditioned space and the external environment or other part of the building as the case may be.

Paragraph (b)(iii) only applies to a building that is not a dwelling, where the other part of the building is used for a purpose that is not similar or identical to the purpose for which the conditioned space is used.

Regulation 3: Meaning of building work

One of the first tasks with respect to any proposal is to establish whether it is **building work** and hence requires a submission under the Building Regulations. Building work is defined as:

(a) the erection or extension of a building;

(b) the provision or extension of a controlled service or fitting in or in connection with a building;

(c) the material alteration of a building, or a controlled service or fitting;

(d) work required by Regulation 6 (requirements relating to material change of use);

(e) the insertion of insulating material into the cavity wall of a building;

(f) work involving the underpinning of a building;
(g) work required by Regulation 22 (requirements relating to a change of energy status);
(h) work required by Regulation 23 (requirements relating to thermal elements);
(i) work required by Regulation 28 (consequential improvements to energy performance).

Where building work under Regulation 3, (g), (h) or (i) does not constitute a material alteration then that work need only comply with the applicable requirements of Part L.

A **material alteration** occurs if the work, or any part of it, would at any stage result in a non-compliance, where it previously complied or, if it did not comply with a relevant requirement, by it becoming **more unsatisfactory**. Only Requirements A, B1, B3, B4, B5 and M, relating to structure, fire safety and disabled access, are relevant to a material alteration. Examples would include an opening in a load-bearing wall, erection of new internal partitions giving rise to increased means of escape travel distances, or the removal of a disabled toilet or access ramp.

Regulation 4: Requirements relating to building work

Building work, as established above, must be carried out in accordance with the relevant requirements listed in Schedule 1. These include the structural stability of the building, means of escape and fire safety, resistance to moisture, ventilation arrangements, drainage, stair design, thermal insulation and facilities for disabled people. To comply with one requirement should not cause a non-compliance with another requirement.

Where compliance with the relevant requirements of Schedule 1 was not originally shown, the building work shall not make the situation **more unsatisfactory** than it was before the work was carried out.

Regulation 5: Meaning of material change of use

From a building regulation point of view a **material change of use** only occurs where the use of a building is changed to:

(a) a dwelling;
(b) contain a flat;
(c) a hotel or boarding house;
(d) an institution;
(e) a public building (as defined in Regulation 2);
(f) a building no longer exempt under Classes I–VI of Schedule 2;
(g) a building containing a greater or lesser number of dwellings;

(h) a building now containing a room for residential purposes;
(i) a building containing a greater or lesser number of rooms for residential purposes; or
(j) a shop.

Regulation 6: Requirements relating to material change of use

Depending on the actual change of use that will take place, certain **relevant requirements** of Schedule 1 will need to be applied:

* B1, B2, B3, B4(2), B5, C2(c), F1, G1, G3(1)–(3), G4, G5, G6, H1, H6, J1–J4, L1 and P1 for all cases;
* A1–A3 for cases (c), (d), (e) and (f) described in Regulation 5 above (e.g. a barn conversion to a hotel);
* B4(1) in the case of a building over 15 m in height;
* C2 for case (a) (i.e. a dwelling);
* E1–E3 for cases (a), (b), (c), (g) (e.g. a flat conversion) (h) or (i);
* E4 for case (e) (where now a school);
* M1 for cases (c), (d), (e) or (j) (e.g. office conversion to a shop); and
* C1(2) for cases 5 (a), (b), (c), (d), (g), (h), (i) (f – where new residential accommodation);
* G2, G3(4) for case 5(a).

Regulation 7: Materials and workmanship

The functionally written regulation states that building work complying with the relevant requirements of Schedule 1 shall be carried out with adequate and **proper materials** and in a **workmanlike manner**. Regulation 7 is supported by its own Approved Document, the contents of which are discussed later in the book.

Regulation 8: Limitation on requirements

This important regulation clarifies that compliance with Parts A to D, F to K (except paragraphs H2 and J6), [N in Wales only] and P is limited to secure **reasonable** standards of health and safety for persons in or about buildings (and others who may be affected by buildings or matters connected with them). Therefore only a reasonable duty of care rests with the designer and/or builder as far as building regulations are concerned. This also applies from the viewpoint of the Local Authority, who need only establish compliance with the requirements as limited above. This was the subject of case law, *Murphy* v *Brentwood District Council* [1988], where losses of a financial nature were not recognized.

Regulation 9: Exempt buildings and work

Schedule 2 lists the classes of buildings and work that are exempt from the application of the building regulation requirements:

Class I	Buildings controlled under other legislation (i.e. the Manufacture and Storage of Explosives Regulations 2005, Nuclear Installations Act 1965 and the Ancient Monuments and Archaeological Areas Act 1979).
Class II	A detached building not frequented by people where isolated by at least 1.5 times the building height from a controlled building or boundary.
Class III	Greenhouses used for domestic purposes. Agricultural buildings, including a building principally for the keeping of animals (e.g. a stable), where it is not a dwelling, is isolated by at least 1.5 times the building height from sleeping accommodation, and is provided with a fire exit within 30 m. A greenhouse or agricultural building would not be exempt if it was principally used for retailing, packing or exhibiting.
Class IV	Temporary buildings (e.g. mobile homes), not intended to remain erected for more than 28 days.
Class V	Ancillary buildings (i.e. site buildings containing no sleeping accommodation).
Class VI	Small detached buildings containing no sleeping accommodation are exempt up to 15 m² in floor area or up to 30 m² where the building is at least 1 m from a boundary or is constructed of substantially non-combustible materials. A nuclear fallout shelter is also exempt subject to a maximum 30 m² floor area and it being isolated from another building or structure by the depth of the excavation plus 1 m.
Class VII	Extensions up to 30 m² (i.e. conservatory, porch, covered yard, covered way or carport with at least two open sides), although conservatories and porches incorporating glazing should satisfy the requirements of Part K4, K5.1–5.4 [N in Wales only].

Note

Part P applies to electrical installations in greenhouses and small detached buildings.

Regulation 10: The Metropolitan Police Authority

The Metropolitan Police Authority gains exemption from the procedural requirements as a public body.

Regulation 11: Power to dispense with or relax requirements

All the requirements contained within Schedule 1 are written in a functional form and should not therefore be relaxed. The **Dispensation** of a particular requirement may be reasonable in the circumstances whereupon application can be made to the Local Authority. If refused, a right of appeal exists, under Section 39 of the Building Act 1984, to the Department for Communities and Local Government. A relaxation under the Act does not apply in the case of Regulation 23, 25A, 25B, 26, 29 and 29A.

A **Determination** can also be requested from the Department of the Environment, under Section 16(10) of the Building Act 1984, so as to resolve a question between the controlling authority and the applicant. Under this procedure the work or element of work in question should not have commenced at the time of making the application for a determination. A fee for the application is applicable, which stands at half the plan fee (excluding the VAT) with minimum and maximum limits of £100 and £1000 respectively, all as stated in the Building (Local Authority Charges) Regulations 2010.

Regulation 12: Giving of a building notice or deposit of plans

A person intending to carry out building work must make a submission in one of two forms where the Local Authority system of building control is to be utilized (please also refer to Figure 1.1, which also illustrates the alternative route using an **Approved Inspector**):

(a) give to the Local Authority a **building notice** (Regulation 13); or
(b) deposit **full plans** (Regulation 14); this will be necessary for a building to which the Regulatory Reform (Fire Safety) Order 2005 applies (i.e. the majority of commercial premises); erection of a building fronting on to a private street; and if H4 imposes a requirement.

A submission will not be necessary for the types of work listed, and carried out by a person described, in Schedules 3 and 4. The content of Schedule 3 is contained within Table 1.1 and Table 1.2 for Schedule 4. This regulation does not apply where Regulation 19 of the Building (Approved Inspectors, etc.) Regulations (Local Authority powers in relation to partly completed work) applies. For emergency repairs a building notice can be submitted as soon as reasonably practicable after commencement of such work.

Regulation 13: Particulars and plans where a building notice is given

The giving of a building notice represents a simple method of notifying the Local Authority that building work or a change of use is proposed. It requires the submission of the following information:

- the name and address of the person intending to carry out the work and signed by him or on his behalf;
- a statement that the notice is given in accordance with Regulation 12(2)(a);
- a description of the proposal; and
- the location and proposed use of the building.

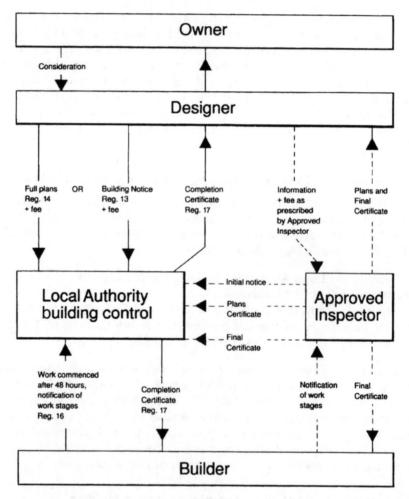

Figure 1.1 Building regulation submission flowchart.

Note
Dashed line route represents the alternative method of gaining design and work approval utilizing an **Approved Inspector**. For specific guidance, reference should be made to the Building (Approved Inspectors, etc.) Regulations 2010 or an Approved Inspector should be consulted (e.g. BRCS (Building Control) Limited).

Table 1.1 Schedule 3: Regulations 12(6)(a) and 20(1)

Self-certification schemes and exemptions from requirement to give building notice or deposit full plans

Type of work	Person carrying out work
1. Installation of a heat-producing gas appliance. This paragraph does not apply to the provision of a masonry chimney.	A person, or an employee of a person, who is a member of a class of persons approved in accordance with regulation 3 of the Gas Safety (Installation and Use) Regulations 1998.
2. Installation of— (a) an oil-fired combustion appliance; or (b) oil storage tanks and the pipes connecting them to combustion appliances. This paragraph does not apply to the provision of a masonry chimney.	A person registered in respect of that type of work by Association of Plumbing and Heating Contractors (Certification) Limited, Benchmark Certification Limited, Building Engineering Services Competence Assessment Limited, Certsure LLP, HETAS Limited, NAPIT Registration Limited, Oil Firing Technical Association Limited, or Stroma Certification Limited.
3. Installation of a solid fuel-burning combustion appliance other than a biomass appliance. This paragraph does not apply to the provision of a masonry chimney.	A person registered in respect of that type of work by Association of Plumbing and Heating Contractors (Certification) Limited, Building Engineering Services Competence Assessment Limited, Certsure LLP, HETAS Limited, or NAPIT Registration Limited.
4. Installation of a heating or hot water system, or its associated controls.	A person, or an employee of a person, who is a member of a class of persons approved in accordance with regulation 3 of the Gas Safety (Installation and Use) Regulations 1998, or a person registered in respect of that type of work by Association of Plumbing and Heating Contractors (Certification) Limited, Benchmark Certification Limited, Building Engineering Services Competence Assessment Limited, Certsure LLP, HETAS Limited, NAPIT Registration Limited, Oil Firing Technical Association Limited, or Stroma Certification Limited.
5. Installation of a mechanical ventilation or air conditioning system or associated controls, in a building other than a dwelling, that does not involve work on a system shared with parts of the building occupied separately.	A person registered in respect of that type of work by Building Engineering Services Competence Assessment Limited, Certsure LLP, NAPIT Registration Limited, or Stroma Certification Limited.
6. Installation of an air conditioning or ventilation system in a dwelling, that does not involve work on a system shared with other dwellings.	A person registered in respect of that type of work by Building Engineering Services Competence Assessment Limited, Certsure LLP, NAPIT Registration Limited, or Stroma Certification Limited.
7. Installation of an energy efficient lighting system or electric heating system, or associated electrical controls, in buildings other than dwellings.	A person registered in respect of that type of work by Building Engineering Services Competence Assessment Limited, Certsure LLP, NAPIT Registration Limited, or Stroma Certification Limited.

continued ...

Table 1.1 continued . . .

Self-certification schemes and exemptions from requirement to give building notice or deposit full plans

Type of work	Person carrying out work
8. Installation of fixed low or extra-low voltage electrical installations in dwellings.	A person registered in respect of that type of work by BSI Assurance UK Limited, Benchmark Certification Limited, Building Engineering Services Competence Assessment Limited, Certsure LLP, NAPIT Registration Limited, Oil Firing Technical Association Limited, or Stroma Certification Limited.
9. Installation of fixed low or extra-low voltage electrical installations in dwellings, as a necessary adjunct to or arising out of other work being carried out by the registered person.	A person registered in respect of that type of work by Association of Plumbing and Heating Contractors (Certification) Limited, Benchmark Certification Limited, Building Engineering Services Competence Assessment Limited, Certsure LLP, NAPIT Registration Limited, or Stroma Certification Limited.
10. Installation, as a replacement, of a window, rooflight, roof window or door in an existing dwelling.	A person registered in respect of that type of work by BM Trada Certification Limited, BSI Assurance UK Limited, CERTASS Limited, Certsure LLP, by Fensa Limited under the Fenestration Self-Assessment Scheme, by NAPIT Registration Limited, Network VEKA Limited, or Stroma Certification Limited.
11. Installation, as a replacement of a window, rooflight, roof window or door in an existing building other than a dwelling. This paragraph does not apply to glass which is load bearing or structural or which forms part of a glazed curtain walling or a revolving door.	A person registered in respect of that type of work by BM Trada Certification Limited, CERTASS Limited, Certsure LLP, by Fensa Limited under the Fenestration Self-Assessment Scheme, or by Stroma Certification Limited.
12. Installation of a sanitary convenience, sink, washbasin, bidet, fixed bath, shower or bathroom in a dwelling, that does not involve work on shared or underground drainage.	A person registered in respect of that type of work by Association of Plumbing and Heating Contractors (Certification) Limited, Benchmark Certification Limited, Building Engineering Services Competence Assessment Limited, Certsure LLP, HETAS Limited, NAPIT Registration Limited, or Stroma Certification Limited.
13. Installation of a wholesome cold water supply or a softened wholesome cold water supply.	A person registered in respect of that type of work by Association of Plumbing and Heating Contractors (Certification) Limited, Benchmark Certification Limited, Building Engineering Services Competence Assessment Limited, Certsure LLP, HETAS Limited, or NAPIT Registration Limited.
14. Installation of a supply of non-wholesome water to a sanitary convenience fitted with a flushing device, that does not involve work on shared or underground drainage.	A person registered in respect of that type of work by Association of Plumbing and Heating Contractors (Certification) Limited, Benchmark Certification Limited, Building Engineering Services Competence Assessment Limited, Certsure LLP, HETAS Limited, NAPIT

continued . . .

Table 1.1 continued . . .

Self-certification schemes and exemptions from requirement to give building notice or deposit full plans

Type of work	Person carrying out work
	Registration Limited, or Stroma Certification Limited.
15. Installation in a building of a system to produce electricity, heat or cooling— (a) by microgeneration; or (b) for renewable sources (as defined in Directive 2009/28/EC of the European Parliament and of the Council on the promotion of the use of energy from renewable sources).	A person registered in respect of that type of work by Association of Plumbing and Heating Contractors (Certification) Limited, BRE Global Limited, Benchmark Certification Limited, Building Engineering Services Competence Assessment Limited, Certsure LLP, HETAS Limited, NAPIT Registration Limited, Oil Firing Technical Association Limited, or Stroma Certification Limited.
16. Installation, as a replacement, of the covering of a pitched or flat roof and work carried out by the registered person as a necessary adjunct to that installation. This paragraph does not apply to the installation of solar panels.	A person registered in respect of that type of work by NAPIT Registration Limited or the National Federation of Roofing Contractors Limited.
17. Insertion of insulating material into the cavity walls of an existing building.	A person registered in respect of that type of work by Cavity Insulation Guarantee Agency Limited under the Cavity Wall Insulation Self Certification Scheme, by Certsure LLP, NAPIT Registration Limited, or Stroma Certification Limited.
18. Installation of insulating material to the internal walls of a building not including the installation of flexible thermal linings.	A person registered in respect of that type of work by British Board of Agrément, CERTASS Limited, Certsure LLP, NAPIT Registration Limited, or Stroma Certification Limited.
19. Installation of insulating material to the external walls of a building, not including installation of demountable-clad buildings.	A person registered in respect of that type of work by British Board of Agrément, CERTASS Limited, Certsure LLP, NAPIT Registration Limited, or Stroma Certification Limited.
20. Installation of insulating material to the external and internal walls of a building ("hybrid insulation"), not including installation of demountable-clad buildings, and not including the installation of flexible thermal linings.	A person registered in respect of that type of work by British Board of Agrément, Certsure LLP, or NAPIT Registration Limited.

Third party certification schemes and exemptions from requirement to give building notice or deposit full plans

Type of work	Person inspecting work
Electrical installations in dwellings.	A person registered in respect of that type of work by BSI Assurance UK Limited, NAPIT Registration Limited, or Stroma Certification Limited.

Note: Schemes may not apply in Wales

Table 1.2 Work that need not be notified to building control body

Work that need not be notified to building control body, as contained in Schedule 4 of the 2010 Building Regulations, consisting of:

- existing fixed building service, where testing/adjustment not possible, the replacement of any part (not a combustion appliance, other than a fixed flueless gas cooker), adding an output device or adding a control device (excluding a fixed or external lighting system);
- a self-contained building service (not a combustion appliance, excluding a fixed or external lighting system and a mechanical ventilation appliance located in a room containing an open-flued combustion appliance);
- replacing an external door (door and frame not more than 50% glazed);
- providing fixed internal lighting to no more than 100 m², in existing buildings other than dwellings;
- replacement of a WC (using no more water), washbasin, sink, bidet, fixed bath, shower, rainwater gutter or downpipe (excluding underground drainage);
- existing cold water supply, the replacement of any part, adding an output device or adding a control device;
- providing a hot water storage system where storage vessel capacity does not exceed 15 litres; and
- installation of thermal insulation to roof or loft space, and suspended timber floors.

For an erection or extension of a building the following are also needed:

- a 1:1250 or greater scale plan showing the size, position and curtilage boundaries of the building and other buildings and streets within that curtilage;
- the number of storeys (including basement storeys) in the building;
- drainage provisions; and
- steps to be taken to ensure compliance with any local enactment.

To enable the Local Authority to verify compliance with the regulations they may also request in writing the submission of further plans within a specified time period. These could include additional structural calculations, detailed floor layout drawings for means of escape purposes, or the specification of a disabled person passenger lift.

It should be noted that a building notice, or the plans submitted with it, is not treated as a formal submission for building regulation approval under Section 16 of the Building Act 1984. The choice could therefore be made to submit a full plans application, which would give the benefit of an Approval Notice and a Completion Certificate once the plans and work on site had been found to comply with the regulations. Note that unless the work has commenced a building notice becomes of no effect after three years.

Regulation 14: Full plans

As previously stated, the deposit of a full plans application offers the dual benefits of an Approval Notice and Completion Certificate. A certain degree of information is required to enable the authority to check full compliance with the Building Regulations:

- the name and address of the person intending to carry out the work and signed by him or on his behalf;
- a statement that the notice is given in accordance with Regulation 12(2)(b);
- a description of the proposal;
- the location and proposed use of the building;
- a 1:1250 or greater scale plan showing the size, position and curtilage boundaries of the building and other buildings and streets within that curtilage;
- the number of storeys (including basement storeys) in the building;
- drainage provisions and precautions to be taken if building is over a public sewer (reference Sections 21 and 18 of the Building Act, 1984 respectively);
- steps to be taken to ensure compliance with any local enactment;
- a statement to confirm if the Regulatory Reform (Fire Safety) Order 2005 applies to the building.

In addition to all of the above, 'any other plans which are necessary to show that the work would comply with these regulations' shall also be deposited. In this regard it is very important to identify the extent of the information that the Local Authority will need to verify compliance with the Regulations. For example, reams of builder's specification notes or working details can be very time-consuming to produce and are unlikely to be required. However, it will be necessary to provide detailed floor lay-outs, plans and sections to an appropriate scale and specific details where compliance with a particular regulation needs to be shown, e.g. damp-proof course arrangements at an external wall/roof abutment or the provision of sound insulation around an internal soil and vent pipe.

The submission of full plans shall be in duplicate with two additional copies provided where Part B: Fire safety applies, but with the exception of houses and flats. This allows the Local Authority to consult the local fire authority, and is explained later in the book.

Once full plans have been deposited and acknowledged by the Local Authority they must approve or reject them within five weeks, or two months if an extension of time is agreed. Any rejection notice must state where the plans do not conform to the Regulations or where additional information is necessary.

A further benefit of the full plans procedure is that plans may be passed conditionally by one of these two methods:

- where the plans show a contravention the Local Authority may approve them subject to the necessary correction being made, e.g. a damp-proof course not indicated or a particular door not specified as a fire door; and/or
- the Local Authority may approve the plans subject to further specific plans being deposited at a later date, normally before that specific element of work has commenced, e.g. design and details of timber, trussed rafters or the design and specification of an active fire safety feature. This method can enable the plans to be dealt with in stages.

In both cases the Local Authority is not obliged to use them and the applicant must give written agreement.

Regulation 15: Consultation with sewerage undertaker

This regulation applies where full plans have been deposited with the Local Authority and paragraph H4 of Schedule 1 imposes requirements. [Similar provisions apply under The Building (Approved Inspectors, etc.) Regulations.]

The Local Authority shall consult, with sufficient plans, the sewerage undertaker, as soon as practicable after the plans have been deposited, and before issuing any completion certificate. The sewerage undertaker is allowed 15 days to respond and the Local Authority shall have regard to the views expressed.

Regulation 16: Notice of commencement and completion of certain stages of work

This regulation states where notice needs to be given to the Local Authority (Regulation 12) by the person carrying out building work and before that work has commenced:

- Two days' notice for commencement (i.e. any period of 48 hours commencing at midnight, excluding weekends and public/bank holidays). Note that the full plans submission or building notice needs to have been with the authority for this time before work can commence.
- The Local Authority may then give notice, by way of an inspection framework, of the specific stages they intend to inspect, where notification would be required and for how long the work must remain uncovered. This would be based on their risk assessment of a breach of the Regulations occurring if an inspection was not undertaken.
- Five days' notice for completion and occupation.

The manner in which notice is given is not stated, but could include writing, e-mail or telephone. Where notice has not been given and the work has been covered up, the Local Authority can require the work to be cut into, laid open or pulled down where necessary to establish compliance. For example, a trial hole and core sample may be deemed necessary to ascertain the depth and quality of foundation concrete already cast and not seen by the Local Authority building control officer.

It is important to note that building work may be commenced, constructed and completed without the benefit of a Building Regulation approval or having to construct the building work in accordance with the approved plan. This is subject to giving the notices described above and constructing the building work itself so as to comply with the relevant requirements of Schedule 1 to the Building Regulations 2010 (as amended). This regulation does not apply for work supervised by an Approved Inspector.

Regulation 17: Completion certificates

The Local Authority has an obligation to issue a completion certificate within eight weeks when notified of completion.

The Local Authority will need to take **reasonable steps** to ensure that the building works show compliance with the relevant requirements of Schedule 1. A certificate issued under this regulation can be regarded as evidence, but not conclusively, that the relevant requirements specified in the certificate have been complied with.

Regulation 17A: Certificate for building occupied before work is completed

The Local Authority also has an obligation to issue a completion certificate within four weeks in respect of part or all of a building where:

- it is to be occupied before the work is completed; and
- the Regulatory Reform (Fire Safety) Order 2005 applies to the building.

The Local Authority will need to take **reasonable steps** to ensure that, regardless of completion of the current building work, those parts of the building to be occupied before completion comply with regulation 38 and Part B of Schedule 1.

Regulation 18: Unauthorized building work

To regularize a situation where unauthorized building work has taken place, on or after 11 November 1985, and a submission has not been deposited, the owner or applicant may request in writing that the Local Authority

issue a **Regularization Certificate**. The submission of the following information will be necessary:

- a statement that the application is made in accordance with Regulation 21;
- a description of the unauthorized work;
- a plan of the unauthorized work, if reasonably practicable; and
- a plan of any additional work, if reasonably practicable, where necessary
  to enable compliance to be shown with the relevant requirements
  applicable at the time of construction.

To enable the Local Authority to establish that compliance has been shown
they may require the applicant to take **reasonable steps** in laying open the
unauthorized work. This could include, for example, the excavation of trial
holes to verify foundation depths, the testing of drainage, or the taking of
building material samples for analysis. It may also be necessary to relax or
dispense with a requirement, as described under Regulation 11 above.

Once the Local Authority has taken all reasonable steps, notified the
applicant accordingly and satisfied itself that the unauthorized building
work now complies with the building regulations it *may* issue a Regularization Certificate. A certificate issued under this regulation can be regarded
as evidence, but not conclusively, that the relevant requirements specified
in the certificate have been complied with.

It should be noted that this regulation is without prejudice to any action
the Local Authority may take under Section 36 of the Building Act 1984
relating to the **removal or alteration of offending work**.

Regulation 19: Supervision of building work otherwise than by local authorities

If the building work is to be supervised by an **Approved Inspector** (or public
body), where an initial notice and final certificate are given, then Regulations
12, 16, 17, 17A, 20, 27, 29, 37, 41, 42, 43, 44, 45 and 46 shall not apply.
Where the services of an Approved Inspector are to be utilized then reference
should be made to the Building (Approved Inspectors, etc.) Regulations
2010; please also refer to Figure 1.1.

Regulation 20: Provisions applicable to self-certification schemes

For building work as outlined in Schedule 3 the Local Authority is author-
ized to accept and store, as evidence that the requirements of Regulations
4 and 7 have been satisfied, a certificate to that effect by the person carrying
out the building work. This certificate, and notice to the Local Authority,

shall be deposited in not more than 30 days after that work has been completed. Note that this does not apply to work contained in Schedule 4 of the regulations (please refer to Table 16.1 on p. 279).

Regulation 20A: Provisions applicable to third party certification schemes

Similar to the provisions of Regulation 20 this regulation has been introduced to cover building work, as described at the end of Table 1.1. At present this method of showing compliance only applies to electrical work in dwellings.

Regulation 21: Application of energy efficiency requirements

The energy efficiency requirements basically apply to the erection of any building, or extension, if it is a roofed construction having walls and uses energy to condition the indoor climate. The only buildings excluded are those which are:

- listed, within a conservation area or are included in the schedule of monuments;
- used primarily or solely as places of worship;
- temporary (planned time of use of 2 years or less), industrial sites, workshops and non-residential agricultural buildings with low energy demand; and
- stand-alone with a floor area less than 50 m^2 and not used as a dwelling.

Regulation 22: Requirements relating to a change to energy status

The applicable requirements of Part L, Schedule 1, apply where there is change to a building's energy status. For example, where an unheated warehouse is to become heated office accommodation.

Regulation 23: Requirements for the renovation or replacement of thermal elements

This requirement specifically relates to the renovation or replacement of an individual thermal element, e.g. a roof or wall panel, either as a major renovation or to more than 50% of the element's surface area. Subject to it being technically, functionally and economically feasible the whole of the thermal element must be renovated or replaced to comply with Part L, paragraph L1(a)(i).

Regulation 24: Methodology of calculation and expression of energy performance of buildings

The Secretary of State shall approve:

- a methodology of calculation of the energy performance of buildings, including methods for calculating asset ratings and operational ratings of buildings; and
- ways in which the energy performance of buildings, as calculated in accordance with the methodology, shall be expressed.

In this regulation **asset rating** means a numerical indicator of the amount of energy estimated to meet the different needs associated with a standardized use of the building; and **operational rating** means a numerical indicator of the amount of energy consumed during the occupation of a building over a period of time.

Regulation 25: Minimum energy performance requirements for buildings

The Secretary of State shall approve minimum energy performance require-ments for new buildings, in the form of target CO_2 emission rates (target fabric energy efficiency rates for dwellings), which shall be set in accordance with the methodology approved pursuant to Regulation 24.

Regulation 25A: Consideration of high-efficiency alternative systems for new buildings

An analysis shall be undertaken for any new building, and made available to the Local Authority before it starts on site, to take into account the technical, environmental and economic feasibility of using high-efficiency alternative systems, e.g.

- decentralised energy supply systems using energy form renewable sources;
- cogeneration;
- district or block heating or cooling, particularly using energy from renewable sources; and
- heat pumps.

Regulation 25B: Nearly zero-energy requirements for new buildings

From 2021, where any new building is erected, it must be a nearly zero-energy building.

[Regulation 25C: New buildings: minimum energy performance requirements]

[To comply with Regulation 24 in Wales a methodology approved by the Welsh Ministers can be used.]

Regulation 26: CO_2 emission rates for new buildings

Where a building is erected, it shall not exceed the target CO_2 emission rate for the building that has been approved pursuant to Regulation 25.

Regulation 26A: Fabric energy efficiency rates for new dwellings

Where a new dwelling is erected, it shall not exceed the target fabric energy efficiency rate for the dwelling that has been approved pursuant to Regulation 25.

[Regulation 26A: Primary energy consumption rates for new buildings]

[Regulation 26B: Fabric performance values for new dwellings]

[In Wales where a new building or dwelling is erected, it shall not exceed the target primary energy consumption rate or target fabric performance value approved pursuant to Regulation 25C.]

Regulation 27: CO_2 emission rate calculations

To verify compliance with Regulation 26 the person carrying out the work shall provide to the Local Authority a notice that specifies: the target CO_2 emission rate; and the calculated CO_2 emission rate for the building as constructed. A Local Authority can accept a certificate to that effect by a person who is an accredited energy assessor.

Regulation 27A: Fabric energy efficiency rate calculations

To verify compliance with Regulation 26A the person carrying out the work shall provide to the Local Authority a notice that specifies; the target fabric energy efficiency rate for the dwelling; the calculated fabric energy efficiency rate for the dwelling as designed; and a list of specifications. A local authority can accept a certificate to that effect by a person who is an accredited energy assessor.

[Regulation 27A: Primary energy consumption rate calculations]

[Regulation 27B: Fabric performance values calculations]

[To verify compliance with Regulation 26A in Wales reference is made to a target fabric performance value for dwellings and a target primary energy consumption rate for all other buildings.]

Regulation 28: Consequential improvements to energy performance

Where building work to an existing building (with a total useful floor area over 1000 m²) consists of an extension; the initial provision of any fixed building services; or an increase to the installed capacity of any fixed building services; it shall comply with the requirements of Part L of Schedule 1. This would not apply if the work is not technically, functionally and economically feasible.

Regulation 29: Energy performance certificates

This regulation applies where:

- a building is erected; or
- a building is modified so that it has a greater or lesser number of parts designed or altered for separate use than it previously had, where the modification includes the provision or extension of any of the fixed services for heating, hot water, air conditioning or mechanical ventilation.

The person carrying out the work shall give an **energy performance certificate** for the building to the owner of the building, and give to the Local Authority notice to that effect. Similar provisions apply under the Building Approved (Inspectors, etc.) Regulations. The **energy performance certificate** shall be given not later than the date on which the notice required by Regulation 30(4) is required to be given to the Local Authority, where that regulation applies; or not later than five days after the work has been completed, in any other case.

The **energy performance certificate** must:

- be accompanied by a recommendation report, issued by the energy assessor;
- express the asset rating of the building in a way approved by the Secretary of State under Regulation 24;
- include a reference value such as a current legal standard or benchmark;
- be issued by an energy assessor who is accredited to produce energy performance certificates for the category of building to which the certificate relates; and

- include the registered reference number; the address of the building; an estimate of the total useful floor area of the building; the name of the energy assessor who issued it; the date on which it was issued; and the name of the approved accreditation scheme of which the energy assessor is a member.

Regulation 29A: Recommendation reports

Any recommendation report made by an **energy assessor** must include:

- cost-effective measures under a major renovation or for individual elements of the building envelope or fixed building services; and
- where any owner or tenant can obtain more detailed information and the steps to be taken to improve the building energy efficiency.

Regulation 30: Energy assessors

An **energy assessor** must be a member of an accreditation scheme approved by the Secretary of State. Where an energy assessor issues an energy performance certificate and recommendation report he must ensure they are entered onto the relevant register maintained by the Secretary of State in accordance with Regulation 37 of the Energy Performance of Buildings (Certificates and Inspections) (England and Wales) Regulations 2012 before he gives them to the person who requested that he issue them.

Regulation 31: Related party disclosures

An energy assessor must include a declaration of any personal or business relationships in the energy performance certificate.

Regulation 32: Duty of care

An energy assessor must carry out energy assessments with reasonable care and skill. Any cause of action arising in relation to the duty imposed is deemed not to be an action founded on tort for the purposes of the Limitation Act 1980(24).

Regulation 33: Right to copy documents

Any person may, for the purpose of complying with any duty imposed by these Regulations or the Energy Performance of Buildings (Certificates and Inspections) (England and Wales) Regulations 2007, copy or issue a copy of any document produced by an energy assessor.

Regulation 34: Application of building regulations to educational buildings, buildings of statutory undertakers and Crown Buildings

This regulation seeks compliance for these buildings with Regulations 11(1), 21, 23, 25, 25A, 25B 26 (CO_2 emission rate) and 29.

Regulation 35: Interpretation of Part 6

Definitions are given for the terms outlined in the regulations above.

Regulation 36: Water efficiency of new dwellings

The potential consumption of wholesome water should not exceed 125 litres per person per day for each dwelling, see Section G2 of Chapter 9.

Regulation 37: Wholesome water consumption calculation

A notice needs to be served to the Building Control Body to verify the water consumption under Regulation 36 above, see Section G2 of Chapter 9.

[Regulation 37A: Provision of automatic fire suppression systems]

[This regulation only applies in Wales to building work consisting of the erection or material change of use of a building in relation to:

- care homes (as section 3 of the Care Standards Act 2000);
- rooms for residential purposes other than those in a hostel, hotel, prison or young offender institution and a hospital;
- dwelling-houses and flats, from 1 January 2016.]

[Regulation 37B: Requirements of automatic fire suppression systems]

[The design of an automatic fire suppression system, to accord with Regulation 37A, should follow the guidance within the Welsh version of Approved Document B.]

Regulation 38: Fire safety information

To assist the **responsible person** to operate and maintain the erected **relevant building** or extension (or building after a **relevant change of use**) with reasonable safety, appropriate **fire safety information** must be provided to them. This relates to the design and construction, services, fittings and equipment, all as outlined in Chapter 4, Appendix G, and should be provided at or prior to completion or occupation.

Regulation 39: Information about ventilation

This applies where paragraph F1(1) imposes a requirement. Within five days of completion sufficient information is to be provided to the owner to operate the ventilation system.

Regulation 40: Information about use of fuel and power

This applies where paragraph L1 imposes a requirement. Within five days of completion sufficient information is to be provided to the owner to allow the efficient operation of the building and its fixed building services.

Regulation 41: Sound insulation testing

The introduction of sound testing, where paragraph E1 imposes a requirement, requires the person carrying out the work to:

- ensure appropriate testing, and recording of results, in accordance with a procedure approved by the Secretary of State; and
- provide a copy of the results, with the notice under Regulation 16(4), to the Local Authority.

Similar provisions apply under The Building (Approved Inspectors, etc.) Regulations.

Sound testing is not, however, required for houses or flats where the adopted design details are one or more of those approved by Robust Details Limited. The Local Authority must be notified prior to commencement of work. This notification must specify the part or parts of the building in respect of which he is using the design detail, the design detail concerned, and the unique number issued by Robust Details Limited. The building work must then comply with the notification details.

Regulation 42: Mechanical ventilation air flow rate testing

This applies to the creation of a new dwelling where paragraph F1(1) imposes a requirement to test, in accordance with a procedure approved by the Secretary of State, the mechanical ventilation air flow rate, with results to be submitted to the Local Authority.

Regulation 43: Pressure testing

Air tightness testing, where paragraph L1(a)(i) imposes a requirement, requires the person carrying out the work to:

- ensure appropriate testing, and recording of results, in accordance with a procedure approved by the Secretary of State; and
- provide a copy of results to the Local Authority. (A Local Authority can accept a certificate to that effect by a person who is registered by

the British Institute of Non-destructive Testing or the Air Tightness and Testing and Measurement Association in respect of pressure testing for the air tightness of buildings.)

Regulation 44: Commissioning

Fixed building services, where paragraphs F1(2) and L1(b) impose a requirement, require the person carrying out the work to give notice to the Local Authority that they have been commissioned, where possible, in accordance with a procedure approved by the Secretary of State.

Similar provisions apply under the Building (Approved Inspectors, etc.) Regulations for Regulations 27, 43 and 44.

Regulation 45: Testing of building work

The Local Authority may make such tests of any building work to establish compliance with Regulation 7 or any of the applicable requirements contained in Schedule 1. (Similar provisions apply under the Building (Approved Inspectors, etc.) Regulations.)

Regulation 46: Sampling of material

The Local Authority itself may take samples for testing so as to establish compliance with the regulations.

Regulation 47: Contravention of certain regulations not to be an offence

Namely that an offence under Section 35 of the Building Act 1984 is not applicable for Regulations 17, 17A, 25A, 27, 27A, [Welsh Regulations 27A and 27B], 29, 37, 41, 42, 43 and 44; and Regulations 23, 25B and 26, applicable to Crown buildings.

Regulation 48: Electronic service of documents

Section 94A of the Building Act 1984 now allows building regulation applications, and supporting notices, to be served on the Local Authority electronically.

Regulations 49–53: Transitional provisions

Regulation 54: Revocations

Schedule 3 lists the regulations that are revoked, including the Building Regulations 2000.

Chapter 2

Approved Document to support Regulation 7: Materials and workmanship

---

Regulation 7 states:
Building work shall be carried out:

(a) with adequate and proper materials that:

   (i) are appropriate for the circumstances in which they are used;
   (ii) are adequately mixed or prepared; and
   (iii) are applied, used or fixed so as to perform adequately the functions for which they are designed; and

(b) in a workmanlike manner.

The functionally written regulation is basically saying that building work, to comply with the relevant requirements of Schedule 1, shall be carried out using appropriate workmanship, with materials that are:

- of a suitable nature and quality in relation to the purposes and conditions of their use;
- adequately mixed or prepared; and
- applied, used or fixed so as to perform adequately the functions for which they are intended.

The 2013 edition of this Approved Document acknowledges our relationship with the rest of Europe and the implementation of European Regulation 305/2011/EU-Construction Products Regulation. The Approved Document is split into two sections reflecting the factors listed in the regulation.

Note that the reference to **materials** includes those naturally occurring, e.g. stone, timber and thatch, and products, components, fittings, items of equipment and excavation backfill in connection with building work. The appropriate use of recycled and recyclable materials should now be considered. It is also important to bear in mind the contents of Regulation 8, which clarifies the limited standards required to comply with the Building Regulations and the fact that no continuing control of materials in use is enforced after the work has been completed.

Section 1: Materials

To establish the fitness of a material one of a number of methods can be chosen:

- **CE Marks** - These normally denote compliance with a harmonized European Product Standard or where a European Technical assessment has been obtained by the manufacturer of the product. Reference can be made to the NANDO website at http://ec.europa.eu/enterprise. It is important to check that the material not only has a CE mark but a supporting declaration of performance suitable for the building work in question. This must then be accepted by the building control body. The onus of proof is therefore with the local building control authority (or Approved Inspector), who should notify the trading standards officer in any particular case.
- **British Standards** – An appropriate British Standard or harmonised European Standard may be referred to.
- **Other national and international technical specifications** – Conformity to the national technical specifications of other Member States.
- **Independent certification schemes** - Many schemes exist within the UK, including those accredited by the United Kingdom Accreditation Service (UKAS). A certification body, accredited by a national accreditation body belonging to the European co-operation for Accreditation (EA) can also be relied upon.
- **Tests and calculations** – The UKAS or EA Accreditation Schemes for Testing Laboratories are examples. This would ensure that the tests, calculations or other means are carried out or undertaken in accordance with agreed criteria and can be relied upon.
- **Past experience** – An existing building, where the material in question has been used, may verify its suitability in use and over a reasonable time span.
- **Sampling** – Reference should be made to Regulation 46, which clarifies Local Authority powers to take and sample materials. Similar powers also exist for an Approved Inspector.

Short-lived materials may be regarded as unsuitable, owing to their potential deterioration over a short period of time if not correctly maintained. If accessibility is possible for inspection, maintenance and repair, then a

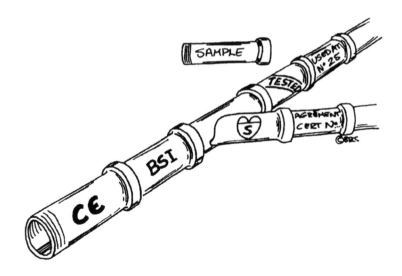

particular material used in a particular location may be acceptable. This would be subject to the consequences of failure not being serious to persons in or about the building, e.g. an external cladding system up to a modest height where cladding panels could be removed for inspection periodically. It may also be acceptable to use a short-lived material that is not accessible but where the risk to health and safety is low, e.g. a single-layer roofing membrane to a storage building.

Materials susceptible to changes in their properties

Examples of such materials have now been removed and reference made to two conditions where materials may be used in building work:

- Residual properties, including structural stability, can be estimated.
- Residual properties adequate for the expected life of the building.

Section 2: Workmanship

To establish the adequacy of a particular method of workmanship a number of methods are available, which are outlined as follows:

- **CE Mark** – The relevant European Technical Assessment or harmonized product standard may specify the workmanship.
- **Standards** – An appropriate British Standard Code of Practice may be referred to, or an equivalent national technical specification of other Member States (e.g. BS 8000: *Workmanship on building sites*).

- **Independent certification schemes** – Many schemes, including those accredited by UKAS and EA will specify workmanship. Competent person self-certification schemes can also be relied upon.
- **Management systems** – Justification of quality may be possible by the utilization of a scheme showing compliance with BS EN/ISO 9000.
- **Past experience** – An existing building, where the workmanship method in question has been used, may verify its suitability to perform the desired function.

- **Tests** – Reference should be made to Regulation 45, which clarifies Local Authority powers to test drainage. In addition testing is required for sound (Regulation 41), air flow (Regulation 42) and pressure (Regulation 43).

The Approved Document concludes with appendices for key terms, standards and documents referred to.

Chapter 3

Approved Document A: Structure

Approved Document A: Structure is the first in a series of documents approved by the Secretary of State to offer **practical guidance** on compliance with the Building Regulations. Approved Document A specifically offers guidance on the functional Requirements A1, A2 and A3 contained in Part A of Schedule 1 to the Regulations.

In general the guidance contained within the Approved Document is relatively straightforward, with reliance placed on the vast range of British Standards and Eurocodes available to the designer. The opening Sections 1–4 can be used to verify compliance with Requirements A1 and A2.

REQUIREMENT A1: LOADING

The building shall be constructed so that the combined dead, imposed and wind loads are sustained and transmitted by it to the ground:

(a) safely; and
(b) without causing such deflection or deformation of any part of the building, or such movement of the ground, as will impair the stability of any part of another building.

In assessing whether a building complies with the above regard shall be had to the imposed and wind loads to which it is likely to be subjected in the ordinary course of its use for the purpose for which it is intended.

REQUIREMENT A2: GROUND MOVEMENT

The building shall be constructed so that ground movement caused by:

(a) swelling, shrinkage or freezing of the subsoil; or
(b) landslip or subsidence (other than subsidence arising from shrinkage), in so far as the risk can be reasonably foreseen,

will not impair the stability of any part of the building.

To take account of potential ground movement, buildings should be constructed to transmit loads safely to the ground and not impair the stability of other buildings. The structural safety of a building therefore depends on the successful interrelationship between design and construction, taking into account these particular aspects:

- loading from dead, imposed and wind loads (taking into account possible dynamic, concentrated and peak load effects);
- properties of materials;
- design analysis;
- details of construction;
- workmanship; and
- safety factors (taking into account all the above aspects).

With these aspects in mind the guidance put forward in Sections 1–4 is now discussed.

Section I: Codes, standards and references

This section lists various codes and standards that may be used for the structural design and construction of *all* buildings (Table 3.1).

Table 3.1 Design codes, standards and corresponding Eurocodes

Design & Loading		BS 6399: Part1: 1996 (dead and imposed loads) BS 6399: Part 2: 1997 (wind loads) BS 6399: Part 3: 1988 (imposed roof loads)	BS EN 1990 and BS EN 1991
Timber		BS 5268: Part 2: 2002 BS 5268: Part 3: 1998 BS 8103: Part 3: 1996	BS EN 1995
Masonry		BS 5628: Part 1: 1992 BS 5628: Part 2: 2000 BS 5628: Part 3: 2001 BS 8103: Part 1: 1995 BS 8103: Part 2: 1996	BS EN 1996
Concrete		BS 8110: Part 1: 1997 BS 8110: Part 2: 1985 BS 8110: Part 3: 1995	BS EN 1992
Steel		BS 5950: Part 1: 2000 BS 5950: Part 2: 2001 BS 5950: Part 3: 1990 BS 5950: Part 4: 1994 BS 5950: Part 5: 1998 BRE Digest 437	BS EN 1993 and BS EN 1994
Aluminium		BS 8118: Part 1: 1991 BS 8118: Part 2: 1991	BS EN 1999
Foundations		BS 8002: 1994 BS 8004: 1986	BS EN 1997

Note
The British Standards listed have now been withdrawn, reference should be made to the corresponding Eurocodes. For a full listing of the 10 Eurocodes, made up of 58 Parts, please visit http://shop.bsigroup.com/Browse-By-Subject/Eurocodes/.

With respect to foundation design particular attention should be paid to:

- the need to research conditions of ground instability, both known or recorded; and
- the availability, from licence holders of the Department of the Environment, of **reviews** containing regional reports of various geotechnical conditions. The Approved Document lists contacts/availability.

Regulations 5 and 6 apply Requirements A1–A3 to certain buildings undergoing a material change of use. To appraise these buildings structurally, reference can be made to the following documents:

- BRE Digest 366: Structural appraisal of existing buildings, including for a material change of use (2012);
- The Institution of Structural Engineers Report *Appraisal of existing structures* (2010).

For information other guidance sources include:

- TRADA publications; and
- National House Building Council Standards.

Section 2: Sizes of structural elements for certain residential buildings and other small buildings of traditional construction

This section opens with a list of definitions, which apply throughout the section:

Buttressing wall – a wall providing full-height lateral support to another wall perpendicular to it.

Dead load – the weight of all walls, permanent partitions, floors, roofs and finishes including services, and all other permanent construction.

Imposed loads – the weight of movable partitions, distributed, concentrated, impact, inertia and snow loads (i.e. due to the proposed occupancy or use) but excluding wind loads.

Pier – a thickened section of wall at intervals to provide lateral support.

Spacing – centre-to-centre longitudinal spacing of adjacent timber members.

Span – centre-to-centre distance between two adjacent supports or bearings (or between faces of bearings where applicable).

Supported wall – a wall provided with lateral support from a buttress wall, pier, chimney, floor or roof members.

Wind load – the load due to the effect of wind pressure or suction.

Section 2A: Basic requirements for stability

- These are to be read in conjunction with Sections 2B and 2C.
- Trussed rafters and traditional roofs, which are not resistant to instability, should be braced in accordance with BS EN 1995-1-1: 2004.
- If a roof is sufficiently braced and anchored, and walls are designed in accordance with Section 2C, there is no need to take account of wind pressure or suction.

Section 2B: Sizes of certain timber members in floors and roofs for dwellings. Areas at risk from house longhorn beetle

This section now directs you to the TRADA publication *Span tables for solid timber members in floors, ceilings and roofs (excluding trussed rafter roofs) for dwellings.* The following is retained for information.

- Limits for dead and imposed loads are given and common species/grades which may be used are listed.
- A series of clear span tables for timber members are given in the TRADA publication; some specimen values are given in Table 3.2 for information.

Table 3.2 Clear spans for certain timber members in single-family houses

Member	Dead load (kN/m^2)	Span (m) for member size		
		50 × 97 50 × 100	50 × 147 50 × 150	50 × 220 50 × 200
Rafter (at 16°)	<1.0	2.18	3.25	–
Purlin	<1.00	–	–	2.25
Ceiling joist	<0.50	1.89	3.19	5.14
Binder	<0.50	–	1.92	2.64
Sheeting purlin	<0.75	1.24	1.85	2.46
Floor joist	<1.25	1.74	2.81	4.07
Flat roof joist	<1.00	1.81	3.04	4.65

Notes

Roof imposed load at 1.00 kN/m²; for imposed snow loads refer to BS EN 1991.

All members at 400 mm centres, 1500 mm centres for purlins and binders. Grade of timber SC3.

- Double joists may be needed where a floor supports a bath.
- Suitable solid or herringbone strutting should be provided at spans over 2.5 m and two rows of strutting over 4.5 m.
- Notches should be no deeper than 0.125 × joist depth, not cut closer to the support than 0.07 × span or further than 0.25 × span.
- Holes should be no greater than 0.25 × joist depth, drilled at neutral axis, spaced at least 3 × hole diameter and located between 0.25 and 0.4 × span from the support.
- Rafters should only have notches for birdsmouthed bearings.

In certain geographical areas of England softwood timber used for roof construction and within the roof void should be treated to prevent infestation by **house longhorn beetle**. The risk areas are as follows:

- Boroughs of Bracknell Forest (parishes of Sandhurst and Crowthorne), Elmbridge, Spelthorne, Surrey Heath, Rushmoor (Farnborough), Woking.
- Districts of Hart (parishes of Mawley and Yateley) and Runnymede.

Section 2C: Thickness of walls in certain small buildings

- This section applies to residential buildings up to three storeys high and small single-storey non-residential buildings and annexes; size limitations and conditions for use of the guidance are stated in the Approved Document.
- The thickness of coursed brickwork or blockwork solid walls (paragraph 2C6 of the Approved Document refers) should be at least 1/16 of the storey height and accord with guidelines in Table 3.3.
- Uncoursed stone, flint, etc. walls should be at least 1.33 times the thickness determined by paragraph 2C6 of the Approved Document.
- Coursed brickwork or blockwork cavity walls should have at least 90 mm thick leaves, 50 mm cavity, and the combined width + 10 mm should be not less than the thickness determined by paragraph 2C6 above.
- Maximum spacing of cavity wall ties: 900 mm (horizontal) × 450 mm (vertical) spacings for 50–75 mm cavity; see also Table 5 of the Approved Document for the selection of wall tie length.
- Heights of buildings are related to slope of site and exposure.
- Wall panels require end restraint by way of a full-height buttressing wall, pier or chimney.
- Design criteria and limitations of openings, recesses, overhangs and chases are stated in the Approved Document.

Table 3.3 Solid external and compartment wall thickness

Maximum wall panel dimensions (m)	Minimum wall thickness (mm)
	190 for full height
	290 for one storey, 190 for remainder
	290 for two storeys, 190 for remainder

- Wall panels require lateral restraint to restrict movement of the wall at right angles to its plane; consideration will therefore need to be given to strapping details at floor and roof levels.
- Reduced requirements apply to small buildings with single leaf walls, including the need for buttressing walls or piers at 3 m centres.
- For further detailed guidance reference should be made to the Approved Document, including Diagram 6 (revised wind speed map) and Diagram 7 (maximum height of buildings).

Section 2D: Height of chimneys

- Masonry used for the construction of chimneys to have a density of at least 1500 kg/m^3.
- Maximum height (at highest roof intersection and including any pot) to be at least 4.5 × W, where W is the smallest horizontal dimension.

Section 2E: Plain concrete strip foundations

The Approved Document provides rules for the design of foundations, including concrete mix, minimum widths and foundation dimensions. Table 3.4 gives minimum strip foundation widths. Minimum depths are 0.45 m to avoid frost action, and 0.75 m, 0.9 m or 1.0 m for low, medium and high shrinkage clay soils respectively.

Table 3.4 Minimum strip foundation widths

Type/condition of subsoil	Minimum width (mm) for wall loading of					
	20	30	40	50	60	70 kN/linear metre
Rock	- - - - - - - - - width of wall - - - - - - - -					
Compact gravel or sand	250	300	400	500	600	650
Stiff clay or sandy clay	250	300	400	500	600	650
Firm clay or sandy clay	300	350	450	600	750	850
Loose sand, silty sand or clayey sand	400	600	- - - to be calculated - - -			
Soft silt, clay, sandy clay or silty clay	450	650	- - - to be calculated - - -			
Very soft silt, clay, sandy clay or silty clay	600	850	- - - to be calculated - - -			

Section 3: Wall cladding

The section gives guidance on the support and fixing of external wall cladding, and relates to all forms of cladding, i.e. stone, concrete, glass and curtain walling.

To meet Requirement A1 the wall cladding should:

- safely sustain and transmit loadings to supporting structure;
- be securely fixed to give both vertical support and lateral restraint;
- make allowance for differential movement;
- use durable materials, noting degree of accessibility; and
- take account of pedestrian, vehicle and crowd loadings.

Section 3 goes on to outline a technical approach, making reference to relevant British Standards and further guidance, including: Structural use of glass in buildings ISE 1999; Aspects of Cladding ISE 1995; BS 8297; BS 8298; BS 5080; and CIRIA Reports.

Section 4: Roof covering

Materials used to cover roofs should basically sustain the concentrated imposed loads outlined in BS EN 1991.

The recovering of a roof is classified as a **material alteration** under Regulation 3. A procedure should then be followed comprising:

- a comparison between the proposed roof loading and the existing covering to establish if substantially higher or lighter loads will apply (i.e. more than 15%);

- an inspection of the existing roof structure to check whether it is capable of sustaining the increased loads and has adequate vertical restraint against wind uplift if a lighter covering and/or new under-lay is proposed; and
- the adoption of appropriate strengthening measures, for example, the replacement of defective elements, provision of additional structural members and strapping provisions to include resistance to uplift.

Section 5: Reducing the sensitivity of the building to disproportionate collapse in the event of an accident

REQUIREMENT A3: DISPROPORTIONATE COLLAPSE

The building shall be constructed so that in the event of an accident the building will not suffer collapse to an extent disproportionate to the cause.

This requirement followed the Ronan Point disaster, in which an accidental explosion caused the partial collapse of a tower block containing flats.

The application of this requirement is *not* limited to buildings of five or more storeys. Instead, the building should be sufficiently robust to sustain a limited extent of damage or failure without collapse, depending on the consequence class of the building. Table 3.5 clarifies each building class and the strategy to adopt.

As an **alternative approach** reference is made to the following reports: *Guidance on robustness and provision against accidental actions* 1999; *Proposed revised guidance on meeting compliance with the requirements of Building Regulation Part A3* (both available at www.planningportal. gov.uk); and the *Practical guide to structural robustness and disproportionate collapse in buildings* 2010 (ISE). Other guidance sources include: *NHBC Guidance on Disproportionate Collapse:* updated May 2011; and *Masonry design for disproportionate collapse requirements under Regulation A3 of the Building Regulations (England & Wales)* (BDA/AACPA/CBA).

The Approved Document concludes with a list of standards referred to.

Table 3.5 Building classes and design strategies

Conse-quence class	Building type/occupancy	Design strategy
I	Houses <4 storeys; agricultural buildings; buildings into which people rarely go (unless part of the building is closer to another building, or area where people do go, by a distance of 1.5 times the building height).	Design to Approved Document guidance or that within Section 1.
2A	Five-storey single occupancy houses; hotels, flats, apartments and other residential buildings, and offices <4 storeys; industrial buildings and shops (<2000 m²) <3 storeys; single-storey educational buildings; and all buildings <2 storeys to which members of the public are admitted (<2000 m² at each storey).	Class 1 measures and provide effective horizontal ties, *or* effective anchorage of suspended floors to walls.
2B	Hotels, flats, apartments and other residential buildings, and offices >4 storeys but <15 storeys; educational buildings >1 storey but <15 storeys; shops >3 storeys but <15 storeys; hospitals <3 storeys; all buildings to which members of the public are admitted, floor areas >2000 m² but <5000 m² at each storey; and car parking <6 storeys.	Class 1 measures and provide effective horizontal ties, *and* effective vertical ties, *or* check that upon the notional removal of each supporting column and each beam supporting one or more columns, or any nominal length of load-bearing wall (one at a time in each storey of the building), the building remains stable (floor area at risk of collapse not to exceed 15% of the floor area of that storey or 70 m², and does not extend further than the immediate adjacent storeys). Where extent of damage upon removal is exceeded then such elements should be designed as **key elements** (sustain applied load of 34 kN/m²).
3	All buildings > limits in Classes 2A and 2B; grandstands accommodating >5000 spectators; and buildings containing hazardous substances and/or processes.	Undertake a systematic risk assessment of the building, taking into account all the normal hazards that may reasonably be foreseen, together with any abnormal hazards, during the life of the building. The structural form and concept and any protective measures should then be chosen.

Note
For Classes 2A, 2B and 3 refer to BS EN 1990–1996 for design guidance.

Chapter 4

Approved Document B: Fire safety

Introduction

As a consequence of the review of the Building Regulations 1985 undertaken by the Department of the Environment, Approved Document B has changed significantly in a number of key areas, reflecting advances in technology and building form. These areas include the adoption of guidance rather than mandatory rules for means of escape design, a new requirement covering fire service access, and a lessening of standards concerning compartmentation and periods of fire resistance where the benefits of sprinkler systems are further recognized.

The 2006 edition of the Approved Document came into effect in April 2007 and has been split into two volumes. The first deals with dwelling-houses only and the second with buildings other than dwelling-houses. The idea is that Volume 1 will prove less complicated for those owners and designers dealing with domestic housing and extensions. For very large and unusual houses reference would need to be made to Volume 2, which also incorporates fire safety guidance for flats. To avoid repetition this chapter keeps together all fire safety guidance relating to each of the Building Regulation Requirements. To assist eventual owners/occupiers/employers to meet their statutory duties under the Regulatory Reform (Fire Safety) Order 2005 an appendix covering the fire safety information required under Regulation 38 has been added.

The specific aims of Requirements B1, B2, B3, B4 and B5 are highlighted within the general introduction to Approved Document B, and it is compliance with these different aspects of fire safety that must be considered by the designer:

- satisfactory means of warning and escape;
- restriction of fire spread over internal surfaces;
- sufficient stability of the building under fire load;
- sufficient fire separation within the building and between adjoining buildings;
- restriction of fire and smoke spread within concealed spaces;

- restriction of fire spread over the external envelope and from one building to another; and
- satisfactory access for fire appliances and firefighters up to and within the building in the saving of life.

Particular emphasis is placed on their close interrelationship, and where a variance is proposed of one or more of the requirements a trade-off could be possible so as to achieve an acceptable **fire safety package**.

Following on from this, the concept of **fire safety engineering** is introduced so as to achieve a **total fire safety package**. This is especially appropriate for large and complex building forms and existing buildings; smaller projects may also benefit from such an approach. The following specific factors should be assessed:

- risk of fire occurring in the first place;
- fire severity based on the fire load of the building (and its contents);
- structural resistance to fire and smoke spread; and
- potential risk to people in and about the building.

With these factors in mind a range of measures can be applied in varying degrees, including:

- adequate methods of fire prevention;
- swift warning of fire;
- control of smoke movement;
- utilization of active extinguishment methods;
- facilities to assist the fire service;
- staff training in fire safety and evacuation procedures; and
- utilization of continuing control under other legislation.

To quantify these measures a range of fire safety engineering techniques are available. These include zone modelling, computational fluid dynamics, virtual reality simulations, risk analysis, fire threat factors and statistical analysis. Further reference can be made to BS 7974: 2001–2003 *Application of fire safety engineering principles to the design of buildings*. BS 9999: 2008 *Code of practice for fire safety in the design, management and use of buildings* is also a valuable reference and now supersedes the entire BS 5588 series (except Part 1).

Other legislation

In addition to the Building Regulations, consideration may need to be given to the application of other legislation that may impose fire safety requirements, including:

- the Regulatory Reform (Fire Safety) Order 2005, which replaced the Fire Precautions Act 1971 and the Fire Precautions (Workplace)

Regulations 1997 and applies to all non-domestic premises, including the common parts of flats and homes in multiple occupation (HMOs);

- the Housing Act 2004, HMOs;
- the Workplace (Health, Safety and Welfare) Regulations 1992, for common parts of flats;
- the Construction (Design and Management) Regulations 2006, for fire safety during construction;
- the Safety at Sports Grounds Act 1975; and
- guidance published by the Environment Agency.

One final point to bear in mind is that compliance with the Building Regulations is primarily for **life safety**, and offers only a reasonable standard of **property protection** in case of fire. It may be necessary to incorporate additional measures subject to clarification from the building insurers. Examples include the provision of fire walls, sprinkler protection and fire-resisting external walls, reference the LPC *Design guide for the fire protection of buildings* and FPA *Design guide*. See also the Crown Fire Standards (PACE) for buildings in the Civil and Defence Estates.

Procedural guidance

The increased scope of requirements contained in the 1992 edition of Approved Document B influenced and changed the responsibilities of the enforcing authorities. For the majority of cases the Building Control Body (approved inspector or Local Authority) will act as the coordinating authority and have primary responsibility for enforcement concerning pre-occupation, with post-occupation enforcement undertaken by the fire authority.

Adequate and meaningful consultation between these parties and in turn with the designer and owner are therefore of the utmost importance. Because of this the then Department of the Environment, Transport and the Regions/ Home Office/National Assembly for Wales published a document entitled *Building regulation and fire safety – procedural guidance*. The latest, 4th Edition of the guide takes into account the Regulatory Reform (Fire Safety) Order 2005 and changes to the Approved Inspector Regulations. The aim of the document is to offer designers, developers, occupiers and owners a procedural guide for all fire safety issues concerning buildings and building work. It should also act as a model for all fire authorities and building control bodies.

The guide gives a step-by-step process, which is summarized in Figure 4.1, highlighting the division of responsibilities.

A number of important items are raised within the guide, and are listed here for information:

- The Building Control Body *must* consult the fire authority where the building is to be put to a use to which the Regulatory Reform (Fire Safety) Order 2005 applies.

- The consultation procedures should ensure that no extra building work is needed after the building is complete and occupied. Note that the statutory bar no longer exists.
- The **responsible person** (as defined in Article 3 of the Regulatory Reform (Fire Safety) Order 2005) is required to undertake a fire risk assessment of the premises once occupied. As part of the consultation process it is useful if the designer produces a preliminary fire risk assessment to assist in the preparation of the fire safety strategy. This, together with the other **fire safety information** required under Regulation 38 (see Appendix G of the Approved Document) can then be passed to the **responsible person** to allow them to undertake the fire risk assessment.

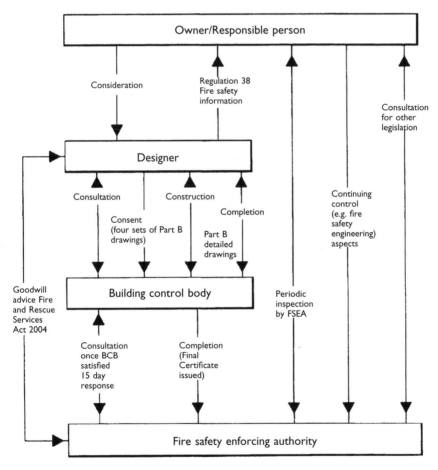

Figure 4.1 Procedural guidance flowchart.

- To control high risk premises the fire authority may serve an **alterations notice** under Article 29 of the Regulatory Reform (Fire Safety) Order 2005.
- Where different building control bodies deal with different parts of a development the applicant must ensure adequate liaison between them.
- Other legislation and local Acts may also require consultation between the Building Control Body and the fire authority.
- From the flowchart it can be seen that before the plans are passed to the fire authority they should show compliance with Part B.
- Where differences of opinion occur between building control and the fire authority they should be dealt with between themselves up to the highest level; if deemed necessary the Department of Communities and Local Government may be consulted. The applicant may also apply to the Secretary of State or the Welsh Ministers for a **determination** of a building regulation or fire safety order issue.
- Note that the building regulations do not address the risk of fire during construction, this is covered by the Construction (Health, Safety and Welfare) Regulations. Enforced by the fire authority for occupied buildings and the Health and Safety Executive when unoccupied.

Purpose group

Before giving consideration to the guidance contained in the Approved Documents the **purpose group(s)** of the proposal must first be established. A purpose group is a method of grading of occupancies based on assumed fire loadings. For this, reference should be made to Table 4.1. A building or compartment may contain more than one purpose group.

In the following situations the different or **ancillary use** should be regarded as a purpose group in its own right:

- a flat or dwelling;
- an area exceeding one-fifth of the total building or compartment floor area where the building or compartment is over 280 m^2; or
- storage in a shop or commercial building (purpose group 4) exceeding one-third of the total building or compartment floor area, where the building or compartment is over 280 m^2.

For buildings with more than one main use, which are not ancillary to each other, each use should be regarded as a purpose group in its own right. Offices over a ground-floor shop are one example. In large buildings where a complex mix of uses occurs consideration should be given to the risk of one use against the other(s). A shopping mall is an example where special measures may be necessary to offset the risk from one shop unit to another (see Section 12).

Table 4.1 Classification of purpose groups[1]

Title	Group	Purpose for which the building or compartment of a building is intended to be used
Residential[2] (dwellings)	1(a)	Flat or maisonette.
	1(b)	Dwelling-house that contains a habitable storey with a floor level which is more than 4.5 m above ground level.
	1(c)	Dwelling-house that does not contain a habitable storey with a floor level which is more than 4.5 m above ground level.
Residential (institutional)	2(a)	Hospital, home, school or other similar establishment used as living accommodation for, or for the treatment, care or maintenance of, persons suffering from disabilities due to illness or old age or other physical or mental incapacity, or under the age of five years, or place of lawful detention where such persons sleep on the premises.
Residential (other)	2(b)	Hotel, boarding house, residential college, hall of residence, hostel and any other residential purpose not described above.
Office	3	Offices or premises used for the purpose of administration, clerical work (including writing, book-keeping, sorting papers, filing, typing, duplicating, machine calculating, drawing and the editorial preparation of matter for publication, police and fire service work), handling money (including banking and building society work), and communications (including postal, telegraph and radio communications) or radio, television, film, audio or video recording, or performance [not open to the public] and their control.
Shop and commercial	4	Shops or premises used for a retail trade or business (including the sale to members of the public of food or drink for immediate consumption and retail by auction, self-selection and over-the-counter wholesale trading, the business of lending books or periodicals for gain and the business of a barber or hairdresser) and premises to which the public is invited to deliver or collect goods in connection with their hire, repair or other treatment, or (except in the case of repair of motor vehicles) where they themselves may carry out such repairs or other treatments.

Table 4.1 (Continued)

Title	Group	Purpose for which the building or compartment of a building is intended to be used
Assembly and recreation	5	Place of assembly, entertainment or recreation; including bingo halls, broadcasting, recording and film studios open to the public, casinos, dance halls, entertainment, conference, exhibition and leisure centres; funfairs and amusement arcades; museums and art galleries; non-residential clubs, theatres, cinemas and concert halls; educational establishments, dancing schools, gymnasia, swimming pool buildings, riding schools, skating rinks, sports pavilions, sports stadia; law courts, churches and other buildings of worship, crematoria; libraries open to the public, non-residential day centres, clinics, health centres and surgeries; passenger stations and termini for air, rail, road or sea travel; public toilets, zoos and menageries.
Industrial	6	Factories and other premises used for manufacturing, altering, repairing, cleaning, washing, breaking up, adapting or processing any article, generating power, or slaughtering livestock.
Storage and other non-residential[3]	7(a)	Place for the storage or deposit of goods or materials [other than described under 7(b)] and any building not within any of the purpose groups 1–6.
	7(b)	Car parks designed to admit and accommodate only cars, motorcycles and passenger or light goods vehicles weighing no more than 2500 kg gross.

Notes
1   This table only applies to Part B.
2   Includes any surgeries, consulting rooms, offices or other accommodation, not exceeding 50 m^2 in total, forming part of a dwelling and used by an occupant of the dwelling in a professional or business capacity.
3   A detached garage not more than 40 m^2 in area is included in purpose group 1(c); as is a detached open carport of not more than 40 m^2, or a detached building, which consists of a garage and open carport, where neither the garage nor open carport exceeds 40 m^2 in area.

The principle of providing people with a safe and effective means of escape from the building to a place of safety outside the building applies to all buildings with the exception of prisons provided under Section 33 of the Prisons Act 1952.

The **guidance** for the designer contained within the Approved Documents is aimed at more straightforward projects and hence reference is made to other guidance documents, including: British Standard 5588: Parts 0, 5–11

REQUIREMENT B1: MEANS OF ESCAPE

The building shall be designed and constructed so that there are appropriate provisions for the early warning of fire, and appropriate means of escape in case of fire from the building to a place of safety outside the building capable of being safely and effectively used at all material times.

(now BS 9999), and Health Technical Memorandums 05–02 and 88. Reference can also be made to the *Fire Safety – Risk Assessment* guides published by the DCLG to support the Regulatory Reform (Fire Safety) Order.

With regard to certain assembly buildings, e.g. theatres, lecture halls and stadia, problems arise with respect to fixed seating arrangements. The Approved Document acknowledges this point and makes direct reference to:

- BS 5588: Part 6: 1991 *Code of Practice for places of assembly* (now BS 9999), and
- *Guide to safety at sports grounds* (TSO).

For schools and other education buildings the fire safety objectives of the Department for Education and Employment's Constructional Standards can be achieved by following the guidance in *Building Bulletin 100*. The main recommendations, which were previously contained within the Approved Document, have been retained in this chapter for reference.

Where individual shops are brought together to form a complex or shopping centre the principles for means of escape design vary, and the Approved Document recognizes this fact by referring the designer to:

- BS 5588: Part 10: 1991 *Code of practice for enclosed shopping complexes* (now BS 9999); and
- Building Research Establishment Report 368 *Methodologies for smoke and heat ventilation*.

With respect to means of escape for disabled people, Part M requires the majority of buildings (or parts) to have suitable access facilities. Therefore reasonable means of escape provisions should be made where the guidance, based on BS 5588: Part 8: 1999 *Code of practice for means of escape for disabled people*, is now contained within the Approved Documents.

B1 is split into two sections for **dwelling-houses**, contained in Volume 1, and five sections for **buildings other than dwelling-houses**, outlined in Volume 2. To avoid duplication of guidance within this chapter, Sections 1 and 2 include the Approved Document guidance for dwelling-houses.

Section I: Fire alarm and fire detection systems

Requirement B1 makes specific reference to the early warning of fire and the appropriate provisions for fire alarm systems.

Suitable provisions for dwelling-houses and flats are:

A fire detection and fire alarm system in accordance with BS 5839: Part 6: 2004 to at least a Grade D Category LD3 standard. Smoke and heat alarms should be mains-operated to BS EN 14604 or BS 5446: Part 2.

- For **large houses,** over 200 m², adopt Grade B Category LD3 or Grade A Category LD2 where dwelling-house over three storeys.
- Where new habitable rooms, without a final exit, are provided by way of a **material alteration** smoke alarms should be provided in the circulation spaces, e.g. loft conversion.
- For sheltered housing schemes, with a warden or supervisor, a connection to a central monitoring point should be incorporated.
- For student accommodation where up to six students share a self-contained unit the guidance for flats can be used. Where general evacuation is required, e.g. hall of residence, then the guidance for buildings other than flats should be followed.

Buildings other than dwelling-houses and flats should all have arrangements for detecting fire. The specification of the fire alarm/detection system will depend on the building type, purpose group and means of escape strategy.

- Small buildings may only require a shouted warning of 'FIRE' that can be heard by all occupants of the premises.
- Manually operated sounders (handbells) or a simple manual callpoint with a bell and power source may also prove suitable means to ensure compliance.
- All other buildings will require the consideration of an electrically operated fire alarm system to BS 5839: Part 1. Category L for the protection of life (subdivided into L1, L2, L3, L4 and L5), Category P for property protection and Category M, manual alarm system, which should be suitable for the majority of cases (purpose groups 3, 4, 5, 6, 7(a) and 7(b)), i.e. manual break-glass call-points and sufficient sounders.
- Visual fire alarms (flashing beacons) should be included where people with impaired hearing may be isolated, e.g. hotel bedrooms and toilet accommodation.
- Shopping complexes and large assembly buildings may warrant a voice alarm, linked with the public address system, reference BS 5588: Part 6 and Part 10 (BS 9999, Annex D and Annex E).
- Automatic fire detection and alarm systems should be provided for purpose groups 2(a) and 2(b); for example, a hotel would normally require a Category L3 system protecting the escape routes and defined parts of the building, including the bedrooms.

Other circumstances may require automatic fire detection:

- protection of an inner room situation;
- compensation for a trade-off where variance is proposed;
- to operate pressurization systems or automatic door release mechanisms;
- to protect unoccupied areas that could prejudice escape from occupied areas; and
- for phased evacuation of a building.

Fire detection and alarm systems must be properly designed, installed and maintained. The installation and commissioning certification should be issued to the Building Control Body and fire authority.

Section 2: Means of escape from dwelling-houses

The provisions for one- and two-storey houses are straightforward but important:

- All habitable rooms (excluding kitchens) in the upper storey(s), served by a single stair, should have an escape window (or external door) at least 0.33 m², 450 mm high, 450 mm wide and not more than 1100 mm above floor level, or direct access to a **protected stairway**. Note that a link door can be used to avoid escape windows to each room.
- All habitable rooms on the ground storey (excluding kitchens) should open directly onto an entrance hall or other suitable exit or should have an escape window or door as above.
- A sleeping gallery will require an alternative exit, or an escape window (where not more than 4.5 m above ground level), if the gallery overlooks less than 50% of the lower floor level and the distance from any part on the upper level to the head of the access stair exceeds 7.5 m. Also a distance of 3 m should not be exceeded from the foot of the access stair to the room exit, and cooking facilities should be enclosed or remote from the access stair.
- Early fire warning, see Section 1.
- Habitable rooms within basements should have an escape window or door, or be served by a **protected stairway.**

Note that a room is termed an **inner room** where its only escape route is via another room or **access room**. Only acceptable for a kitchen, laundry or utility room, dressing room, bath/shower room, WC, other room on floor up to 4.5 m with an escape window and a sleeping gallery. An open-plan house arrangement with a stair open to a ground-floor room would give rise to this situation. Guidance for the installation of smoke alarms is given in Section 1, including the minimum requirement of one unit per storey, interconnected where there is more than one.

For houses that have one or more floors above 4.5 m, measured from ground level, more stringent provisions apply. This is due to the greater risk associated with escape via high-level windows.

Provisions for three-storey houses, in addition to the above, are:

- the utilization of a **protected stairway** giving directly to a final exit (or via two final exits separated by fire-resisting construction); or
- the separation of the top storey by fire-resisting construction and the provision of an **alternative escape route** leading to its own final exit.

A protected stairway is a minimum half-hour fire-resisting enclosure containing the staircase, having fire doors, designated FD 20, to all rooms (except a bathroom or WC) and discharging to a final exit.

For loft conversions to existing two-storey dwelling-houses a **protected stairway** should be provided, i.e. as if it were a new three-storey house. The main provisions are illustrated in Figure 4.2.

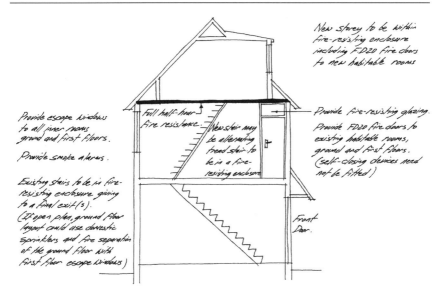

New storey to be within
fire-resisting enclosure
including FD20 fire doors
to new habitable rooms

Provide fire-resisting glazing.
Provide FD20 fire doors to
existing habitable rooms,
ground and first floors.
(self-closing devices need
not be fitted)

Provide escape windows
to all inner rooms
ground and first floors.

Provide smoke alarms.

Existing stairs to be in fire-
resisting enclosure giving
to a final exit(s).
(If open plan, ground floor
layout could use domestic
sprinklers and fire separation
of the ground floor with
first floor escape windows)

Full half-hour
fire resistance.

New stair may
be alternating
tread stair to
be in a fire-
resisting enclosure

Front
Door.

Figure 4.2 Provisions for loft conversion.

For houses of four storeys and above an alternative escape route should be provided to all floor levels over 7.5 m; or a domestic sprinkler system fitted throughout the dwelling-house to BS 9251: 2005.

The remaining provisions for dwelling-houses can be listed as follows:

- Fire doors to protected stairway do not need to be self-closing, only a door leading to an integral garage.
- An escape route to an enclosed yard or garden can be accepted if its depth exceeds the height of the house/rear extension.
- A cavity barrier or fire-resisting ceiling should enclose the head of a protected stairway to a dwelling-house with a floor over 4.5 m.
- External walling, doors and windows within a 1.8 m zone (9 m below stair and 1.1 m above top landing) should offer a minimum half-hour period of fire resistance where adjacent to an external escape stair or route.
- Escape routes over a flat roof or balcony should be within the same building, lead to a storey/final exit and offer minimum half-hour fire-resisting construction within 3 m of the route.
- Air circulation systems within a dwelling-house over 4.5 m must not allow fire and smoke to enter the protected stairway, and thus transfer grilles are to be avoided in the enclosing construction/doors.
- Replacement windows, as controlled fittings, should not adversely effect means of escape from the dwelling-house.

Section 2: Means of escape from flats

As with housing, few provisions are necessary for basement, ground and first-floor storeys, i.e. inner rooms and smoke alarms. It is only with the increase in height that the degree of risk increases, calling into play further detailed provisions. The inherent compartmentation between flats (see also Requirement B3) leads to a low probability of fire spreading from the dwelling of origin. Hence the complete evacuation of the building is not always necessary, and residents can therefore remain within their homes. Satisfactory means of escape provisions must still be provided, however, and these should allow the safe evacuation from the building to a place of ultimate safety, i.e. the outside air at ground level. In the design of such provisions specific consideration needs to be given to the spread of smoke, prevention of fire and fire spread within the common parts of the building, and the fact that reliance should not be placed on external rescue.

In broad terms the Approved Document guidance reflects the principles contained within BS 5588: Part 1: 1990, and indeed reference is made to it for other less common layouts. The Approved Document guidance is also applicable to **houses in multiple occupation** and **sheltered housing schemes**.

Escape routes within each dwelling and escape from the dwelling itself, via any common corridor or stair, to the final exit, represent the two components for satisfactory means of escape design, which can be summarized as follows.

Escape within dwelling

For one- and two-storey flats, where a floor level is not more than 4.5 m above ground level, apply the guidance in Section 2: Means of escape from dwelling-houses.

For floors more than 4.5 m above ground level adopt (a), (b) or (c) for flats and (d), (e), (f) or (g) for multi-storey flats (without their own external entrance):

(a) Provide a **protected entrance** hall, where the maximum **travel distance** within it from a door to any habitable room to the entrance door does not exceed 9 m; or
(b) limit the maximum overall travel distance, from any point in the flat, to the entrance door to 9 m, with the cooking facilities kept remote from the escape route; or
(c) provide an alternative exit from the bedroom accommodation, which itself should be separated from the living accommodation by half-hour fire-resisting construction.
(d) Provide **alternative exits** to each habitable room not located on the entrance floor level; or
(e) provide an **alternative exit** to the floor level (not entrance floor level) utilizing a **protected landing/hall**; or

(f) a **protected stairway** with smoke alarms in all habitable rooms, heat alarm in kitchen (where the vertical distance between the entrance storey floor and the floors above and below it do not exceed 7.5 m); or

(g) a **protected stairway** *and* a domestic sprinkler system to BS 9251: 2005.

Note that an alternative exit should be remote from the main entrance door to the flat and lead to a final exit or common stair via a suitably protected access corridor, common balcony, internal private stair, external stair, common stair or a route over a flat roof (cross-reference Approved Document K).

- Provide *within* each flat early fire warning, see Section 1, including the minimum requirement for multi-storey flats of one unit per storey interconnected.
- Air circulation systems within a flat over 4.5 m must not allow fire and smoke to enter the protected stairway, and thus transfer grilles are to be avoided in the enclosing construction/doors.
- A sleeping gallery will require an alternative exit, or an escape window (where not more than 4.5 m above ground level), if the gallery overlooks less than 50% of the lower floor level and the distance from any part on the upper level to the head of the access stair exceeds 7.5 m. Also a distance of 3 m should not be exceeded from the foot of the access stair to the room exit, and cooking facilities should be enclosed or remote from the access stair.

Escape within common areas

- Each dwelling, on a floor above 4.5 m from the ground level, should have an alternative escape route from the entrance door so that a person can turn and escape away from the fire. This is not necessary where each dwelling has access to a single common stair which is via a **protected lobby** or corridor and the entrance door is within a maximum travel distance of 7.5 m to the stair enclosure (this distance is also applicable within the dead-end portion of a common corridor; see Figure 4.3).
- Where an alternative escape route is available from the entrance door (i.e. to more than one common stair) then the travel distance within the common corridor to a common stair should not exceed 30 m. The corridor in this case should be subdivided by a self-closing fire door (FD 20S designation incorporating smoke seals; see Figure 4.3).
- For small single-stair buildings modified guidance applies, where a maximum travel distance of 4.5 m from the entrance door to the stair enclosure is acceptable or no travel distance limit where each storey only contains two dwellings subject to:

(a) the top floor not exceeding 11 m (from ground level); and

(b) not more than three storeys above ground level; and

(c) the stair not to connect to a covered car park; and

(d) the stair not to serve ancillary accommodation unless separated with ventilated protected lobby/corridor and no dwelling is located on ancillary accommodation floor; and

(e) the stair to be provided with 1.0 m² openable vent at each landing level, or one at head of stair operable from access level.

- Even with the inherent compartmentation provided within flat developments consideration still needs to be given to smoke dispersal within the common escape routes. This is achieved by openable vents or automatic openable vents, minimum free areas required are indicated in Figure 4.3. Smoke vents serving single stair or dead end situations should open automatically, operated by smoke detection in the common corridor or protected lobby. Where the 1.5 m² openable vent cannot discharge to the outside air a vertical smoke shaft, closed at the base and minimum 1.5 m² in area, can be used.

- As an alternative to the natural ventilation provisions described above mechanical ventilation using pressure differentials can be used, reference BS EN 12101–6: 2005.

- The minimum width for a common stair should be 1.1 m if used as a firefighting stair, otherwise the width for everyday use is acceptable.

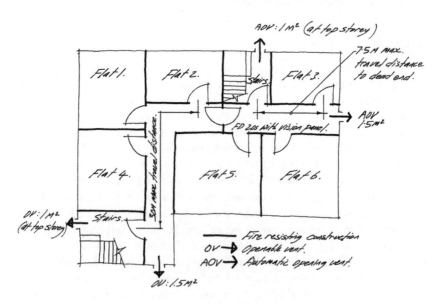

Figure 4.3 Provisions for flats served by two common stairs.

- All common stairs (including any passageway to the final exit) should be constructed as protected stairs and enclosed with fire-resisting construction.
- Two adjoining protected stairways (or exit passageways) should be separated by imperforate fire-resisting construction.
- It is important that a protected stairway contains no potential fire risk; the only allowable exceptions are a lift well or electrical meter(s).
- The stair in a single-stair building should not descend to a basement storey or serve any covered car park, boiler room, fuel storage space or any other ancillary accommodation of similar high fire risk. Other stairs may communicate with these areas subject to the provision of a ventilated protected lobby/corridor at each level.
- Dwellings on not more than three storeys within mixed-use buildings may communicate with other occupancies, where the stairs are separated by protected lobbies. Flats above three storeys should be provided with independent access unless the dwelling is ancillary, has an independent alternative escape route, linked fire alarm system and the stairs separated by protected lobbies.
- Where a flat is also used as a place of work, i.e. live/work unit, then additional fire precautions are required. A maximum 18 m travel distance applies within the work area to any storey or final exit. Windowless accommodation should also be provided with emergency lighting provisions.

[Section 2: Automatic fire suppression systems]

[To show compliance with Regulation 37A in Wales an additional Section 2 has been added within the Welsh version of Approved Document B. The guidance takes into account that provided within the English version of Approved Document B and makes reference to:

- *Sprinklers for Safety: Use and Benefits of Incorporating Sprinklers in Buildings and Structures*, BAFSA 2006 (ISBN: 0 95526 280 1);
- *The Benefits of Automatic Fire Suppression Systems in Residential Care Premises*, BAFSA 2009;
- For dwellings and residential buildings, BS 9251: 2005 *Sprinkler systems for residential and domestic occupancies – Code of practice.*
- The components of the automatic fire suppression systems shall be installed to BS 9252: 2011 *Components for residential sprinkler systems – Specification and test methods for residential sprinklers;* or
- For non-residential buildings BS EN 12845:2009 or an equivalent technical standard.]

Section 3: Horizontal escape for buildings other than flats

As with Section 2 the means of escape design should allow persons to turn their back on a fire and evacuate the building safely. This particular section, applying to all other buildings, considers the portion of the escape route from any point on the floor to the storey exit. This is illustrated in Figure 4.4, where the Boardroom, point A, is regarded as an inner room from which there is only one route of escape. However, alternative escape routes are available from point B.

With the above in mind the first task is to establish the population of any room, tier or storey. If this is not known from the design brief or actual data (average occupant density at a peak trading time of the year) then reference can be made to Table 4.2.

Once the population of any room, tier or storey has been established the minimum number of escape routes/storey exits from that floor level can be found from Table 4.3.

Reference can now be made to Table 4.4, which brings together limitations on **travel distance**. It relates purpose groups to maximum travel distances for escape in one direction only or where escape is available in more than one direction.

Consideration now needs to be given to the adequate width of escape routes and storey exits based on the following:

- 750 mm for up to 60 persons (530 mm between fixed storage racking);
- 850 mm for up to 110 persons;
- 1050 mm for up to 220 persons and 5 mm per person thereafter.

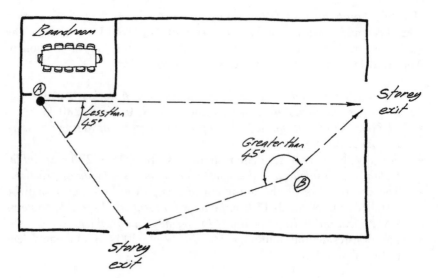

Figure 4.4 Alternative escape routes.

Table 4.2 Floor space factors

Type of accommodation[1,6]	Floor space factor (m²/person)
1. Standing spectator areas, bars without seating and similar refreshment areas.	0.3
2. Amusement arcade, assembly hall (including a general purpose place of assembly), bingo hall, dance floor or hall, club, crush hall, venue for pop concert and similar events.	0.5
3. Concourse, queuing area or shopping mall.[2]	0.7
4. Committee room, common room, conference room, dining room, licensed betting office (public area), lounge or bar (other than in 1 above), meeting room, reading room, restaurant, staff room, waiting room.[3]	1.0
5. Exhibition hall or studio (film, radio, television, recording).	1.5
6. Shop sales area,[4] skating rink.	2.0
7. Art gallery, dormitory, factory production area, museum or workshop.	5.0
8. Office.	6.0
9. Kitchen, library, shop sales area.[5]	7.0
10. Bedroom or study bedroom.	8.0
11. Bedsitting room, billiards or snooker room or hall.	10.0
12. Storage and warehousing.	30.0
13. Car park.	Two persons per parking space

Notes
1   Where accommodation is not directly covered by the descriptions given, a reasonable value based on a similar use may be selected.
2   Refer to Section 4 of BS 5588: Part 10 for detailed guidance on the calculation of occupancy in common public areas in shopping complexes (BS 9999, Annex E).
3   Alternatively the occupant capacity may be taken as the number of fixed seats provided, if the occupants will normally be seated.
4   Shops excluding those under item 9, but including supermarkets and department stores (main sales areas), shops for personal services such as hairdressing and shops for the delivery or collection of goods for cleaning, repair or other treatment or for members of the public themselves carrying out such cleaning, repair or other treatment.
5   Shops (excluding those in covered shopping complexes and including department stores) trading predominantly in furniture, floor coverings, cycles, prams, large domestic appliances or other bulky goods, or trading on a wholesale self-selection basis (cash and carry).
6   If there is to be mixed use, the most onerous factor(s) should be applied.

Table 4.3 Minimum escape routes

Maximum number of persons	Minimum number of escape routes/exits
60	1
600	2
Above 600	3

Note
A single escape route or storey exit may be acceptable where the travel distance in one direction to a storey exit can be complied with and the established population for the area or storey does not exceed 60 persons (or 30 for an institutional use, purpose group 2a).

Table 4.4 Limitations on travel distance

Purpose group	Use of the premises or part of the premises[2]	Maximum travel distance[1] where travel is possible in	
		One direction only (m)	More than one direction (m)
2(a)	Institutional	9	18
2(b)	Other residential:		
	(a) in bedrooms[3]	9	18
	(b) in bedroom corridors	9	35
	(c) elsewhere	18	35
3	Office	18	45
4	Shop and commercial[4]	18	45
5	Assembly and recreation;		
	(a) buildings primarily for the handicapped except schools	9	18
	(b) schools	18	45
	(c) areas with seating in rows	15	32
	(d) elsewhere	18	45
6	Industrial[5]	25	45
7	Storage and other non-residential[5]	25	45
2–7	Place of special fire hazard[6]	9[3]	18[3]
2–7	Plant room or roof top plant:		
	(a) distance within the room	9	35
	(b) escape route not in open air (overall travel distance)	18	45
	(c) escape route in open air (overall travel distance)	60	100

Notes
1 The dimensions in the table are travel distances. If the internal layout of partitions, fittings, etc. is not known when plans are deposited, direct distances may be used for assessment. The direct distance is taken as two-thirds of the travel distance.
2 See also guidance for **small premises**, which has been taken from BS 5588–11 and is now incorporated within the Approved Document.
3 Maximum part of travel distance within the room.
4 Maximum travel distances within shopping malls are given in BS 5588: Part 10: 1991 (BS 9999, Annex E). Guidance on associated smoke control measures is given in a BRE Report 368.
5 In industrial buildings the appropriate travel distance depends on the level of fire risk associated with the processes and materials being used. The dimensions given above assume that the premises will be of 'normal' fire risk, as described in the Home Office Guidance. If the building is high risk, as assessed against the criteria under Note 5 of Table 2 within the Approved Document, then lesser distances of 12 m in one direction and 25 m in more than one direction, would apply.
6 Places of special fire hazard are listed in the definitions in Appendix E.

It is important to take into account the possibility of fire and/or smoke obstructing the use of one of the available exits (Figure 4.5). Therefore where two or more exits exist, the largest one must be discounted so as to determine the adequate width of the remainder. Reference should also be made to Approved Document M concerning adequate widths for disabled access.

The remaining provisions outlined within the Approved Document with respect to horizontal escape can be summarized and illustrated as follows:

- To protect persons in an **inner** room against the occurrence of fire in the **access** room the enclosing walls of the inner room should be stopped a minimum 500 mm short of the ceiling, *or* a door or wall vision panel provided, minimum 0.1 m^2 in area, *or* a suitable automatic fire detection and alarm system so as to alert the occupants of the inner room. This arrangement is only acceptable under certain conditions:

  (a) the inner room population should not exceed 60 persons (30 for purpose group 2a);
  (b) the inner room should not be a bedroom;
  (c) the inner room should be directly off the access room, i.e. not via a corridor;
  (d) the escape route should pass through no more than one access room;
  (e) the appropriate one-direction travel distance should be complied with; and

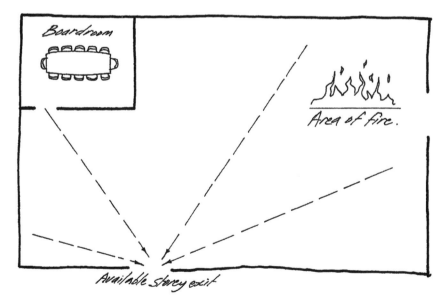

Figure 4.5 Escape widths.

(f) the access room should be under the control of the same occupier and not be a place of special fire hazard (e.g. a switchgear or boiler room).

- Storey exits into a central core arrangement need to be adequately separated to retain protected alternative escape routes (see Diagram 13 of the Approved Document).
- **Open spatial planning** may prejudice means of escape, for example, an open lightwell arrangement. Any escape route should not therefore pass within 4.5 m of the opening.
- Travel via one stairway enclosure to reach another and the use of a stairway as a main circulation route at the same level should both be avoided, unless doors fitted with automatic release mechanisms.
- An ancillary area used for the consumption of food and/or drink by customers should have at least two direct escape routes (avoiding areas of higher fire hazard).
- Escape routes from different **occupancies** (i.e. separate ownerships or tenancies of different organizations) within the same storey should not pass via each other, and any common corridor or circulation space should be constructed as a protected route unless a suitable automatic alarm and detection system is installed throughout the storey.
- Corridors serving bedrooms, corridors common to different occupancies and dead-end corridors should be constructed as **protected corridors** (see Figure 4.6).

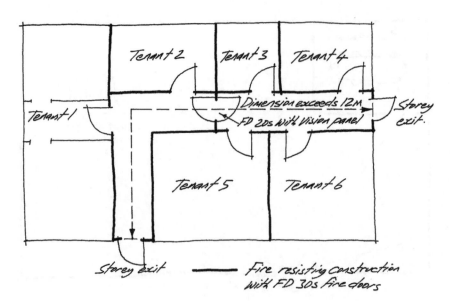

Figure 4.6 Protected corridor.

- Other corridors forming an escape route should have doors to room openings and have partitions taken to floor or ceiling soffit.
- Where corridors exceed 12 m in length and connect two alternative storey exits the corridor should be subdivided at a mid-position with minimum FD 20S self-closing fire doors (see Figure 4.6 and 4.10).
- To protect dead-end lengths of corridor exceeding 4.5 m (and not protected by a pressurization system) they should be separated by self-closing fire doors and fire-resisting construction from any corridor offering two routes of escape or continuation of a route past a storey exit so as to reach another (see Figure 4.7).
- The rules for the protection of external escape routes and those over flat roofs are the same as outlined in Section 3, for dwellings, of the Approved Document. Reference should also be made to Approved Document K: Stairs, ramps and guards.
- For a merging escape flow at a final exit, i.e. a stair and ground floor storey exit, then use the formula $w = ((n/2.5) + (60s))/80$, where w = width of final exit, n = number of persons served by ground floor storey exit, and s = stair width in metres. Note that minimum depth of lobby should be 2 m where n is over 60.

In **small premises** populations are generally low and clear views exist across small undivided floor areas. Reduced provisions, as originally contained in BS 5588–11, can therefore apply. These are subject to: premises

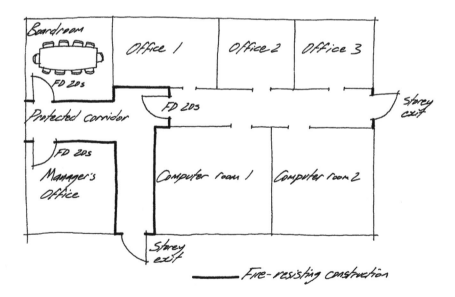

Figure 4.7 Dead-end corridor.

not used for storage/sale of highly flammable liquids or materials; single occupancy on basement/ground/first floor floors only; each floor area not exceeding 280 m²; remotely located kitchens, etc. and store rooms enclosed in fire-resisting construction. Maximum population for a bar or restaurant is 30 persons, or 100 on the ground floor if alternative final exit available.

- A single protected stair, with no ground floor protected lobby, can be used with 18 m travel distances to the storey exits, 27 m allowed on the ground floor.
- For allowable travel distances to **small premises** with open or unprotected stairs refer to guidance in Section 4.

Reference has already been made to Department of Health 'Firecode' documents, which may also be used for non-NHS premises. One of the fundamental principles that is applied to hospitals and other residential care premises is the utilization of **progressive horizontal evacuation**. This avoids the complete evacuation of the patients or residents from the building or any part of it, allowing escape horizontally into an adjoining compartment or place of relative safety. The door(s) in this line of compartmentation can be regarded as a storey exit, where the travel distance limitations should comply with Table 4.4. The Approved Document verifies that this principle may be of value in some other residential buildings, e.g. a residential care home for elderly mentally frail people, and guidance is offered in this regard for the designer, as follows:

- Each storey, used for in-patient care, should be subdivided into a minimum of three **protected areas** within lines of compartmentation.
- The floor area of the adjoining protected area should be sufficient to accommodate the combined population of the protected area at risk and the adjoining protected area itself.
- Each protected area should have a minimum of two independent escape routes. Maximum travel distance from any point to a storey/final exit should not exceed 64 m.
- Escape routes to avoid ancillary accommodation, for example, store rooms, day rooms, kitchens and staff rooms.
- Number of residents' beds depends on staff coverage, maximum 10 single bedrooms in any protected area.
- Bedrooms to be in 30-minute enclosures, free-swing door closers may be used (hold-open devices used for circulation routes).
- Following recent fires in residential care homes the use of sprinkler systems is acknowledged. Fire doors to bedrooms need not be fitted, protected areas can contain more than 10 beds, and bedrooms may contain more than one bed.

Section 4: Vertical escape for buildings other than flats

To evacuate the population of multi-storey buildings from each floor level down to safety at ground level a sufficient number of escape stairs of adequate width is required. The number of stairs is determined by:

- travel distance limitations (Table 4.4);
- the need for separate stairs to serve assembly and recreation or residential purpose groups within a mixed-use building;
- whether a single stair is acceptable for floor levels not more than 11 m above ground level and where a single escape route at each floor level is acceptable (including a basement level); and
- adequate escape width for the established population likely to use the stairs in an emergency subject to the **minimum** values listed in Table 4.5.

To evacuate people who have difficulty in using a stairway, especially those in a wheelchair, consideration must be given to the provision of refuges. These relatively safe waiting areas are normally located within each protected stairway from each storey. An alternative is to utilize a line of compartmentation or an area in the open air safe from the risk of fire with its own means of escape. A refuge within a protected stair, lobby or corridor should:

- have minimum accessible area of 900 mm × 1400 mm;
- not reduce escape route width;
- be clearly identified with blue mandatory 'Refuge – keep clear' sign; and
- have emergency voice communication system linked to control room or to fire alarm panel postion.

To calculate minimum stair widths two strategies exist, depending on the most appropriate method of evacuating the building or part of the building.

Table 4.5 Minimum stair widths

Situation	Maximum number of persons	Minimum stair width (mm)
Institutional building (excluding staff use)	150	1000
Assembly building (exceeding 100 m²)	220	1100
Any other building (occupancy above 50)	Over 220	See Table 4.6 or 4.7
Any other stair	50	800 mm

Table 4.6 Capacity of a stair for basements and simultaneous evacuation of a building

No. of floors served	Maximum number of persons for stair width								
	1000 mm	1100 mm	1200 mm	1300 mm	1400 mm	1500 mm	1600 mm	1700 mm	1800 mm
1	150	220	240	260	280	300	320	340	360
2	190	260	285	310	335	360	385	410	435
3	230	300	330	360	390	420	450	480	510
4	270	340	375	410	445	480	515	550	585
5	310	380	420	460	500	540	580	620	660
6	350	420	465	510	555	600	645	690	735
7	390	460	510	560	610	660	710	760	810
8	430	500	555	610	665	720	775	830	885
9	470	540	600	660	720	780	840	900	960
10	510	580	645	710	775	840	905	970	1035

The **simultaneous evacuation** of the building must be used for basements, buildings containing open spatial planning arrangements, other residential (e.g. hotel), and assembly and recreation buildings where minimum aggregate stair widths are calculated by either:

- reference to Table 4.6 for buildings up to 10 storeys; or
- use of the formula $P = 200w + 50(w - 0.3)(n - 1)$, where P = number of persons accommodated, w = width in metres, and n = number of storeys.

Phased evacuation is the second strategy option where the fire floor (i.e. floor of fire origin) and the floor above are evacuated first and thereafter two floors at a time. This method allows less disruption to occupants of the building and gives stairs of reduced width due to the limited number of persons making their escape at any one time. A number of conditions apply:

- stairs to be approached via protected lobbies/corridors (except a top storey containing plant rooms);
- each floor to be constructed as a compartment floor;
- building to be protected throughout with a sprinkler system (to BS 5306: Part 2, including life safety provisions where applicable) if a floor is above 30 m;
- an automatic alarm and detection system to be installed to at least L3 standard, BS 5839: Part 1, and an internal speech communication system to allow conversation links from fire brigade access level to fire warden at each floor level;
- consideration must be given to entering firefighters on persons escaping. For very tall buildings over 45 m high, passive measures, beyond

Table 4.7 Minimum aggregate width of stairs designed for phased evacuation

Maximum no. of persons per storey	Stair width (mm)
100	1000
120	1100
130	1200
140	1300
150	1400
160	1500
170	1600
180	1700
190	1800

management techniques, may be required, for example the discounting of a protected stairway.

Phased evacuation may be used in high buildings (over 30 m) and large complex buildings. Minimum widths are calculated by either:

- reference to Table 4.7; or
- use of the formula $[(P \times 10) - 100]$ mm where P = number of persons accommodated on the most heavily occupied storey.

Both methods assume that one stair in a multi-stair building will not be available in an emergency, owing to its being smoke-logged or otherwise affected by fire. Each stair should therefore be **discounted** in turn so as to verify that the remainder have the necessary capacity. This discounting rule does not, however, apply where all escape stairs are approached at each floor level (except the top storey) via a protected lobby, *or* all the stairways are protected by a smoke control pressurization system in accordance with BS EN 12101–6: 2005. Note that a storey exit still needs to be discounted. In addition, a stair in excess of 30 m should not exceed 1400 mm in width or alternatively be provided with a central handrail for an 1800 mm minimum width.

The protection of escape stairs, including any passageway leading to the final exit, should follow established guidelines, which can be summarized as follows:

- Provide a protected enclosure to required period of fire resistance, unless an **accommodation stair** (i.e. not a protected stair) can form part of the internal escape route.
- A protected lobby or protected corridor is deemed necessary for: single-stair buildings (with more than one floor above or below the ground storey); buildings designed for phased evacuation; buildings where the

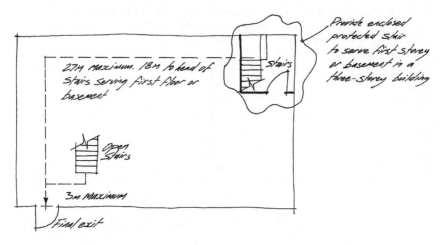

Figure 4.8 Provisions for small two/three storey premises.

stair serves a storey above 18 m; a firefighting shaft; and where a stair needs to be separated from a special fire hazard (smoke ventilation by way of 0.4 m² permanent vent or mechanical smoke control required).

- Adjoining protected stairs should be separated by imperforate construction.
- It is particularly important to restrict potential sources of fire within protected stairs, especially those serving single-stair buildings. The allowable limitations are sanitary accommodation or washrooms (incorporating gas water heaters and sanitary towel incinerators only), a lift well (unless it is a firefighting stair), a reception desk or enquiry office up to 10 m² in area (or equivalent fire risk) but not in a single-stair building, and finally a cupboard enclosed with fire-resisting construction, again not allowable in a single-stair building.
- The protection for external walls of protected stairs and external stairs should follow the guidance in Section 5, for basement stairs refer to Section 2.
- For **small premises,** two/three-storey, not exceeding 90 m² floor area per storey and not containing a restaurant or bar, an open stairway can be used. Please refer to Figure 4.8, travel distances are outlined in Table 4.4.

Section 5: General provisions

This section deals with the construction and protection of escape routes and service installations. It is important that it should be read in conjunction with Sections 2–4.

To safeguard escape routes during the evacuation of a building they should be protected to a minimum 30 minute standard of fire resistance. Consideration also needs to be given to Requirements B3, B5 and other sections of B1 where greater periods of fire resistance may be needed (see Table 4.9). Where glazed elements are incorporated within fire-resisting construction, including doors, they should offer the required period of fire resistance in terms of both integrity and insulation. Where the glazing does not offer the insulation criteria, e.g. Georgian wired glass, then strict limits apply to areas of use, as outlined in Table A4 of the Approved Document. This is because fire presents a radiation hazard for persons trying to escape past glazing with no insulation properties: hence the use of an 1100 mm minimum height (in certain circumstances) above floor level to allow escape under this glazing. Note that additional guidance on the safety of glazing is given in Approved Document N: Materials and protection.

Escape routes should also be constructed to offer a minimum 2 m headroom with no projections below this height, with the exception of door frames. The flooring chosen for all escape routes should be safe in use and minimize slipperiness when wet.

Doors on escape routes represent one of the most important aspects of means of escape design, in terms of both fire resistance and mode of operation. The following is a checklist of the items to consider, together with the contents of Table B1 of the Approved Document:

- Fire doors need to offer the required period of fire resistance and satisfy the test criteria outlined in Appendix B and Table B1 of the Approved Document.
- Generally, all fire doors forming part of an enclosure to common escape routes, protected corridors, lobbies and stairways need to prevent smoke leakage at ambient temperatures, denoted by the suffix (S) in the fire door designation.
- All doors on escape routes should not be fitted with a lock, latch or bolt. Simple fastenings operated from the escape side of the door may be used where only one mechanism needs to be manipulated without the use of a key. The use of 'panic bolts' or similar devices, in assembly or commercial uses for example, is allowable. Additional security locks may be acceptable, when the building is empty, subject to suitable management control.
- Doors should be hung to open in the direction of escape and must be so hung where the population served exceeds 60 persons.
- The swing of a door should be at least 90°, be clear of any change of floor level (except a threshold or single step), not reduce the effective width of a landing or corridor and incorporate a vision panel if hung to swing both ways.

- Revolving and automatic doors or turnstiles used for escape purposes should fail safe, in the open position, in an emergency. If this is not possible adjoining swing doors of sufficient width should be provided.
- Roller shutters should be capable of being opened and closed manually by firefighters. Those across an escape route should only drop on activation of a fusible link or heat detector in close proximity of shutter. Note that certain fire shutters partially descend to form a smoke reservoir in the early stages of a fire.

Turning to the construction of escape stairs, these should utilize **materials of limited combustibility** in the following locations:

- a single-stair building (excluding purpose groups 1a and 3 up to three storeys);
- a basement storey (excluding a private stair within a maisonette);
- a storey more than 18 m above ground or access level;
- an external stair serving a storey more than 6 m above ground level;
- a firefighting stair.

Note that combustible materials may be added to the top surface of these stairs (excluding firefighting stairs).

The escape route may utilize helical and spiral stairs, subject to accordance with BS 5395 *Stairs, ladders and walkways*: Part 2, or a fixed ladder where it is not practical to provide a conventional stair, e.g. to serve a plant room or tank room. Fixed ladders should not be used by members of the public or be constructed of combustible materials. Ramps forming part of an escape route should comply with Approved Document M, and the pitch of a floor or tier slope, e.g. within a theatre, should not exceed 35°. Reference should also be made to Approved Document K: Stairs, ramps and guards.

The remaining general provisions can be summarized as follows:

- Single steps should be prominently marked, those at a door threshold are acceptable, unless required for wheelchair access/escape.
- In certain projecting or recessed protected stairway locations the safe use of the stair may be threatened by fire emanating from adjoining accommodation. In these cases the adjoining external wall construction within 1.8 m should be fire-resisting.
- External walling, doors and windows within a 1.8 m zone (9 m below stair and 1.1 m above top landing) should offer a minimum half-hour period of fire resistance where adjacent to an external escape stair or route.
- Final exits should have a minimum width of the escape routes they serve; should be sited to achieve swift evacuation of the building and

to avoid further threat from fire and smoke; should offer direct access to a street, passageway, walkway or open space; should be well-defined, and guarded where necessary.

- For guidance on escape routes over flat roofs please refer to Section 2.
- Each doorway, exit and route provided for means of escape, except in flats and for exits in normal use, should be provided with suitably located and conspicuous fire exit signage in accordance with BS 5499: Part 1: 1984 or Health and Safety (Safety Signs and Signals) Regulations 1996, e.g. European running man pictogram. Note that additional signage and notices may be necessary to comply with other legislation.
- Protected power circuits should ensure a continuing power supply, where necessary, in the event of fire; minimum PH 30 classification for cabling in accordance with BS EN 50200: 2006.
- Passenger lifts are not usually used for means of escape purposes. However, **evacuation lifts**, suitably sited and protected, can be utilized for the evacuation of disabled persons. For specific guidance reference should be made to BS 5588: Part 8 (and Part 5 and BS 9999, Section 5).
- Lift installations can influence escape routes and hence:

  (a) lift wells should be contained within the protected stair enclosure or enclosed with fire-resisting construction if means of escape is prejudiced;
  (b) lift wells serving different compartments should be regarded and constructed as protected shafts;
  (c) wall climber lifts may penetrate a smoke reservoir, e.g. within an atrium, where its integrity should be preserved;
  (d) in basements and enclosed car parks and where the lift serves areas of high fire risk *and* dwellings or sleeping accommodation then the lift should be approached via a protected lobby (or corridor);
  (e) a lift in a single-stair building or a building where a protected stair enclosure terminates at ground level should not descend down to basement level(s);
  (f) lift machine rooms should be sited above the lift well and outside the enclosure of a single-stair building.

- The design of any mechanical ventilation or air-conditioning system should either close down, or direct potentially smoke-laden air away from any escape route, in the event of fire. Specific reference should be made to BS 5588: Part 9: 1989 (BS 5588: Part 6 and BS 5720) (BS 9999, Section 7).
- Storage chambers, chutes and hoppers used for refuse disposal should be in accordance with BS 5906. Note the need to separate refuse chutes and storage rooms within fire-resisting construction and access provided direct from the external air or via a permanently ventilated (0.2 m²) lobby.

- Walk-in store rooms in shops, which are fully enclosed, should be separated from retail areas with fire-resisting construction if they could prejudice escape routes, unless automatic detection or sprinklers are fitted.
- All escape routes should be provided with adequate artificial lighting *and* **escape lighting** in accordance with BS 5266: Part 1: 2005 to the areas listed in Table 4.8.

Table 4.8 Provisions for escape lighting

Purpose group	Areas requiring escape lighting
Residential	All common escape routes (including external), except two-storey flats.
Office, shop and commercial (where public not admitted), industrial, storage, other non-residential	Underground or windowless accommodation, stairways in a central core or serving a floor above 18 m, internal corridor exceeding 30 m and open plan offices exceeding 60 m².
Shop and commercial and car parks (where public admitted)	All escape routes (including external), except small shops (not restaurant or bar) up to three storeys and maximum 280 m² sales floor area.
Assembly and recreation	All escape routes (including external), and accommodation except where open on one side (daylight hours) and schools (during normal hours).
Any purpose group	Toilets over 8 m² (and under it windowless), electrical generator rooms, switch/battery rooms for emergency lighting system and emergency control room.

REQUIREMENT B2: INTERNAL FIRE SPREAD (LININGS)

1. To inhibit the spread of fire within the building, the internal linings shall:

   (a) adequately resist the spread of flame over their surfaces; and
   (b) have, if ignited, either a rate of heat release or a rate of fire growth which is reasonable in the circumstances.

2. In this paragraph 'internal linings' means the materials or products used in lining any partition, wall, ceiling or other internal structure.

Section 6: Wall and ceiling linings

Wall and ceiling linings are not normally the source of fire, although fire spread and rate of fire growth across these surfaces can be crucial. This is especially so within escape routes and circulation areas, where the rapid spread of fire could prevent persons making their escape. The requirement seeks the use of lining materials that have low rates of surface spread of flame and low rates of heat release. It does not, however, address the problem of fumes and smoke generated by such linings. Compliance should be shown with the classifications indicated in Figure 4.9, depending on location. These classifications are based on fire tests in BS 476: Parts 6 and 7, where Class 1 is the highest, although Class 0 offers the best performance, but is not identified in any British Standard test. The European classifications, as described in BSEN 13501–1: 2002, are now included.

Fire spread across the upper surfaces of floors and stairs is not regarded as a problem in the early stages of a fire and is not therefore controlled. The provision, location and specification of furniture and fittings are also

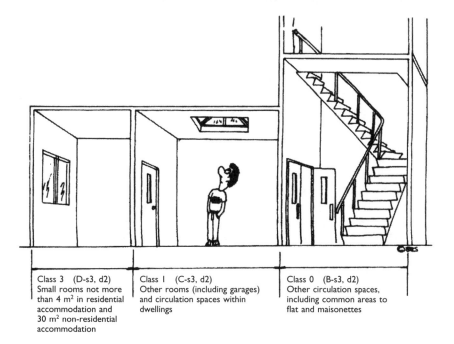

Class 3 (D-s3, d2) Small rooms not more than 4 m² in residential accommodation and 30 m² non-residential accommodation	Class 1 (C-s3, d2) Other rooms (including garages) and circulation spaces within dwellings	Class 0 (B-s3, d2) Other circulation spaces, including common areas to flat and maisonettes

Figure 4.9 Classification of linings.

Notes
A room is taken as an enclosed space that is not used for circulation only, and would include not just conventional rooms but spaces ranging in size from cupboards up to warehouses and auditoria. The European class is indicated in brackets, although this does not automatically equate with the National class.

not controlled since the continuing control of the building and its contents is not possible under building regulations. For certain premises that require a licence, or fall under the Regulatory Reform (Fire Safety) Order 2005, control can be exerted by the fire authority.

The definition of **walls** *includes* the surface of glazing (except door glazing) and any ceiling sloping at more than 70° to the horizontal, but *excludes* doors and doorframes, window and other glazed frames, architraves, cover moulds, picture rails, skirtings and similar narrow members, fireplace surrounds, mantelshelves and fitted furniture. The definition for **ceilings** *includes* the surface of glazing, wall sloping at 70° or less from the horizontal, the underside of a mezzanine or gallery, and the underside of a roof exposed to the room below, but *excludes* trapdoors and frames, window, rooflight and other glazed frames, architraves, cover moulds, picture rails and similar narrow members.

A lower National, or European, class may be adopted for walls in rooms, minimum Class 3, where the wall areas in question do not exceed half of the room floor area subject to a 20 m² maximum for residential buildings and 60 m² for non-residential buildings. A variation is also applicable with regard to rooflights where a lower class may be used, minimum Class 3, subject to the limitations contained in Tables 11 and 18 of the Approved Document.

Special applications are highlighted:

- flexible membrane covering a structure, reference Appendix A of BS 7157;
- PTFE-based materials, reference BRE Report 274.

A **thermoplastic material** is any synthetic polymeric material with a softening point below 200 °C as tested to BS EN ISO 306: 2004 Method A120. These materials are used widely for rooflights, windows, lighting diffusers, suspended or stretched-skin ceilings. A material may not achieve a Class 0–3 designation, and therefore concessions can be made where the thermoplastic material can be classified under one of the following categories, depending on specification and method of fire test:

- TP(a) rigid
- TP(a) flexible
- TP(b).

External windows to rooms only may be glazed with a TP(a) rigid product.

Rooflights to rooms and circulation spaces only may have a lower surface of TP(a) rigid or TP(b) classification subject to the disposition of the rooflights being in accordance with Table 11 and with the guidance concerning external roof surfaces, Section 15 of the Approved Document.

Light diffusers forming part of a ceiling to rooms and circulation spaces only may be unlimited in their extent if classified TP(a) rigid or limited in area and spacing in accordance with Table 11 and Diagrams 27 and 27A of the Approved Document, if TP(b).

A **suspended** or **stretched-skin ceiling** to a room only may be a TP(a) flexible classification where each panel should not exceed 5 m², be supported on all sides and is not a fire resisting ceiling.

REQUIREMENT B3: INTERNAL FIRE SPREAD (STRUCTURE)

1. The building shall be designed and constructed so that, in the event of fire, its stability will be maintained for a reasonable period.
2. A wall common to two or more buildings shall be designed and constructed so that it adequately resists the spread of fire between those buildings. For the purposes of this sub-paragraph a house in a terrace and a semi-detached house are each to be treated as a separate building.
3. Where reasonably necessary, to inhibit the spread of fire within the building, measures shall be taken, to an extent appropriate to the size and intended use of the building, comprising either or both of the following:

    (a) subdivision of the building with fire-resisting construction;
    (b) installation of suitable automatic fire-suppression systems.

4. The building shall be designed and constructed so that the unseen spread of fire and smoke within concealed spaces in its structure and fabric is inhibited.

The wording of Requirement B3 seeks to apply fundamental fire safety principles to buildings, which can be reiterated as follows. Note that B3(3) does not apply to a material alteration to any prison under Section 33 of the Prison Act 1952:

*   provision of fire resistance to elements of structure;
*   subdivision/compartmentation of the building;
*   maintenance of the integrity of elements at openings; and
*   inhibition of fire and smoke spread within concealed spaces.

B3 is split into sections reflecting these principles, with an additional section covering car parks and shopping complexes.

Section 7: Loadbearing elements of structure

The need for a minimum period of fire resistance to the elements of structure of a building is to minimize the risk to:

- occupants remaining in the building;
- people in the vicinity of the building; and
- firefighters occupied in search/rescue operations.

The first task is to establish the **elements of structure** of the building, namely: structural frames, beams, columns, loadbearing walls (internal and external), floor and gallery structures. **Excluded** from this definition are:

- a structure that only supports a roof (including the roof itself) unless it acts as a floor (e.g. a car park or escape route) or if it is essential for the stability of an external wall requiring fire resistance (e.g. by providing restraint);
- the lowest floor of a building; and
- a platform floor (i.e. a raised or access floor containing services).

Reference can now be made to Table 4.9 where the minimum periods of fire resistance are specified. The table brings together *all* purpose groups and relates period of fire resistance to the storey height above and/or below ground level. Single storey is taken under the heading 'not more than 5'.

As with a number of tables in the Approved Document, careful consideration needs to be given to the supporting notes and the contents of Appendix A, which was amended in 2007. The major items are listed for information:

- Depending on the location of the element of structure it will need to satisfy the fire test criteria with regard to **loadbearing capacity (resistance to collapse), integrity** (resistance to fire penetration) and **insulation** (resistance to heat transfer). Table A1 of the Approved Document verifies which criteria apply to a given part of the building and the method of test exposure. For example, a compartment wall separating different occupancies should offer a minimum 60-minute period of fire resistance or as stated by Table 4.9 (whichever is less): the adopted construction will need to satisfy all three test criteria, tested from each side separately. A further example is a cavity barrier, which again should be tested from each side separately but only needs to offer 30 minute integrity and 15 minute insulation. The amended Table A1 in the Approved Document now includes the minimum provisions when tested to the relevant European standard.
- The design and installation of an automatic sprinkler system fitted in accordance with the guidance contained in paragraph 0.16 of the Approved Document can be utilized, for purpose groups 1, 3–7, to give a 30-minute reduction in certain cases and are indeed needed where the building exceeds 30 m to the top storey. The system must cover the complete building and take account of the relevant occupancy and the additional requirements for life safety. Note that high-risk new build properties in Wales, such as care homes, student halls and boarding houses, will require sprinklers. The installation of sprinklers is also proposed for all lower-risk new/converted houses and flats from 1 January 2016.
- Where one element of structure supports or offers stability to another it shall have at least the period of fire resistance of that other element. Variations may be possible, e.g. structure supporting a higher fire risk rooftop plant room.
- An element of structure forming part of one or more compartments should adopt the greater relevant period of fire resistance.
- Basement levels open to the external air on at least one side may adopt the period of fire resistance appropriate to the ground or upper storey.
- Certain elements in a single-storey building may not need any fire resistance, although protection may be necessary to external walls (and supports) to limit the extent of unprotected areas or if the elements support a gallery.

Table 4.9 Minimum periods of fire resistance

Purpose group of building	Minimum periods (min) for elements of structure in a					
	Basement storey[1] including floor over		Ground or upper storey			
	Depth (m) of a lowest basement		Height (m) of top floor above ground, in building or separating part of building			
	More than 10	Not more than 10	Not more than 5	Not more than 18	Not more than 30	More than 30
1. Residential (domestic):						
(a) flats	90	60	30[2]	60[3,7]	90[3]	120[3,4]
(b) and (c) dwelling-houses	Not relevant	30[2]	30[2]	60[8]	Not relevant	Not relevant
2. Residential:						
(a) institutional	90	60	30[2]	60	90	120[5]
(b) other residential	90	60	30[2]	60	90	120[5]
3. Office:						
not sprinklered	90	60	30[2]	60	90	Not permitted
sprinklered	60	60	30[2]	30[2]	60	120[2]
4. Shop and commercial:						
not sprinklered	90	60	60	60	90	Not permitted
sprinklered	60	60	30[2]	60	60	120[5]
5. Assembly and recreation:						
not sprinklered	90	60	60	60	90	Not permitted
sprinklered	60	60	30[2]	60	60	120[2]

6. Industrial:						
not sprinklered	120	90	60	90	120	Not permitted
sprinklered	60	60	30[2]	60	90	120[5]
7. Storage and other non-residential:						
(a) any building or part not described elsewhere:						
not sprinklered	120	90	60	90	120	Not permitted
sprinklered	90	60	30[2]	60	90	120[5]
(b) car park for light vehicles:						
(i) open-sided park	Not applicable	Not applicable	15[2,6]	15[2,6]	15[2,6]	60
(ii) any other park	90	60	30[2]	60	90	120[5]

Notes

1 The floor over a basement (or if there is more than one basement, the floor over the topmost basement) should meet the provisions for the ground and upper storeys if that period is higher.

2 Increased to a minimum of 60 min for compartment walls separating buildings.

3 Reduced to 30 min for any floor within a flat, but not if the floor contributes to the support of the building.

4 Only permitted if sprinklered.

5 Reduced to 90 min for elements not forming part of the structural frame.

6 Increased to 30 min for elements protecting the means of escape.

7 Refer to Section 6 text.

8 30 min in the case of three-storey dwelling houses, increased to 60 min minimum for compartment walls separating buildings.

In the conversion of two-storey single-family dwellings to provide room(s) in the roof it is often found that the existing first-floor construction only offers a **modified half-hour** fire resistance, 15-minute only for integrity and insulation. This is acceptable where only one storey is added, which contains no more than two habitable rooms, does not exceed 50 m^2 in area, separates rooms only and complies with the means of escape guidance in Section 1.

Raised free-standing floors where occupation by persons is limited and does not include members of the public need not have the appropriate period of fire resistance, i.e. to allow an unprotected steel frame. The conditions are that the structure has only one tier and is used for storage purposes only, is not more than 10 m in width or length, does not exceed half of the floor area in which it is located, is open above and below and means of escape comply with Sections 3, 4 and 5. The introduction of automatic detection to the lower level allows floor sizes up to 20 m in width or length. The floor area would not be limited if an automatic sprinkler system is fitted throughout the building.

The conversion of a house into flats can cause problems with regard to the suitability of the existing timber floors. Therefore for buildings up to three storeys a 30-minute period is acceptable; for four storeys and above the full standard should be complied with. Cross-reference may also be needed to Approved Document E: Resistance to the passage of sound.

Section 8: Compartmentation

Subdivision into compartments is a long-established principle and takes into account fire load, ease of evacuation and suppression of fire growth. Compartmentation, utilizing walls and floors of the appropriate fire resistance, should be provided where circumstances dictate. These can be summarized in groups as follows:

- a wall common to two or more buildings;
- the separation of different purpose groups;
- the enclosure of places of special fire hazard;
- the separation of different occupancies within shop and commercial, industrial or storage premises only;
- the separation of a terrace or semi-detached house;
- the separation of an attached/integral domestic garage;
- any wall or floor separating flats and/or other parts of the building;
- a wall enclosing a refuse storage chamber;
- all floors in other residential and institutional purpose groups;
- any wall needed to divide storeys of health care buildings into at least three protected areas;
- any wall needed to subdivide a building with reference to Table 4.10;
- any floor where a storey has a floor over 30 m in height;

Table 4.10 Maximum dimensions of building or compartment (non-residential buildings)

Purpose group of building or part	Height of floor of top storey above ground level (m)	Floor area of any building or any compartment (m²)	One storey in (m²)
		In multi-storey buildings	In single-storey buildings
Office	No limit	No limit	No limit
Assembly and recreation, shop and commercial:			
(a) schools	No limit	800	800
(b) shops – not sprinklered	No limit	2000	2000
shops sprinklered	No limit	4000	No limit
(c) elsewhere – not sprinklered	No limit	2000	No limit
elsewhere – sprinklered	No limit	4000	No limit
Industrial – not sprinklered	Not more than 18	7000	No limit
	More than 18	2000	N/A
Industrial – sprinklered	Not more than 18	14 000	No limit
	More than 18	4000	N/A
		Maximum compartment volume (m³)	Maximum floor area (m²)
		In multi-storey buildings	In single-storey buildings
Storage and other non-residential:			
(a) car park for light vehicles	No limit	No limit	No limit
(b) any other building or part:			
not sprinklered	Not more than 18	20 000	20 000[2]
	More than 18	4000[1]	N/A
sprinklered	Not more than 18	40 000	No limit
	More than 18	8000[1]	N/A

Notes
Certain industrial and storage uses may be controlled by other legislation.
Crown copyright is reproduced with the permission of the Controller of Her Majesty's Stationery Office.
1 Applies to the storeys more than 18 m
2 Maximum height 18 m

- floor over basement and floors within basement if more than 10 m deep;
- separation of shopping complex (see BS 5588: Part 10 (BS 9999, Annex E)).

For a two-storey building, shop, commercial or industrial, the ground storey may be treated as a single-storey building subject to the following:

- upper storey is ancillary and does not exceed 20% of the ground storey area, or 500 m², whichever is less; and
- the upper storey is compartmented and provided with independent means of escape routes.

Compartment walls should generally be taken full height in a continuous vertical plane. Compartment wall or floor junctions need to be carefully considered so as to preserve the integrity of each compartment and to accommodate predicted deflection under fire conditions. A compartment wall, at its junction with a roof, should either continue 375 mm above the roof covering or be fire-stopped to the underside of the covering as detailed in Diagram 30 of the Approved Document, where the roof covering 1.5 m either side of the compartment wall should be designated AA, AB or AC.

The limitation and protection of openings through lines of compartmentation are particularly important so as to restrict the spread of fire. Compartment walls separating buildings or occupancies should be limited to openings for:

- a door for means of escape purposes that has the appropriate fire resistance; and
- the passage of a pipe meeting the provisions of Section 10.

Openings in other compartment walls and floors should be limited to:

- doors that have the appropriate fire resistance;
- the passage of pipes, ventilation ducts, chimneys, appliance ventilation ducts or ducts encasing flue pipes meeting the provisions of Section 10;
- refuse chutes of non-combustible construction;
- atria designed in accordance with BS 5588: Part 7 (BS 9999, Annex B), where atrium breaches compartmentation (reference should also be made to BRE Report 368: *Design methodologies for smoke and heat exhaust ventilation*); and
- protected shafts.

Protected shafts are used to enclose and protect stairs, lifts, escalators, chutes, ducts and pipes passing through lines of compartmentation. The construction and specification of these shafts, which may be horizontal or vertical, should take account of the following items:

- The shaft needs to offer a complete fire barrier to the compartments it serves.
- It should be constructed to offer the appropriate period of fire resistance, noting that this should be achieved from both sides.
- It may include sanitary accommodation or washrooms.
- It may incorporate uninsulated glazed screens.
- It may not contain an oil service pipe (unless it serves a hydraulic lift) or a ventilation duct (unless it pressurizes the stairway) where the shaft also contains stairs or a lift.
- It may contain a natural gas service pipe of screwed steel.
- Adequate ventilation should be provided if the shaft is used to convey flammable gas.
- The external wall element generally need not offer the required period of fire resistance.
- Openings should be protected in line with general compartmentation principles.

Section 9: Concealed spaces (cavities)

The unseen spread of fire and smoke within concealed cavities has resulted in significant damage to buildings and the loss of life. The provisions within this section aim to interrupt cavity firepaths around fire-resisting elements and subdivide extensive cavities by means of cavity barriers. These should offer at least 30-minute fire resistance and be adequately specified and installed to take account of cavity dimension and surrounding construction. The first step is to establish if cavity barriers are needed by referring to Table 4.11.

The provisions in Table 4.12 do not apply to a cavity located in one of the following situations:

- in a wall needing fire resistance only because it is loadbearing;
- in a brick, block or concrete cavity wall, with minimum leaf widths of 75 mm and maximum cavity width of 100 mm, where the cavity is closed at the top of the wall and at the top of each opening;
- in a floor or roof cavity not exceeding a dimension of 30 m above a **fire-resisting ceiling** extending throughout the building or compartment and constructed to accord with Diagram 35 of the Approved Document;
- below a floor next to the ground where the cavity does not exceed 1 m in height or where it is not normally accessible;
- within an underfloor service void;
- behind rain-screen or over-cladding systems subject to the cavity containing no combustible insulation and the provisions of Table 4.11;
- between double-skinned corrugated or profiled **roof sheeting** being of materials of limited combustibility, where the insulation material used is Class 0 or 1 (or Class C-s3, d2 or better European class) and makes contact with both inner and outer sheets;

Table 4.11 Cavity barrier provisions

Consider cavity barrier provisions to the following locations:	Applicable purpose group			
	1a	1b and c	2	3–7
Above a protected stairway in a house of three or more storeys	–	O	–	–
External cavity wall junction with all compartment walls and floors	O	–	O	O
Cavity wall junction with all compartment walls and floors and other fire-resisting barriers	O	–	O	O
Above (and below) a protected escape route	O	–	O	O
Above (and below) bedroom partitions	–	–	O	–
Above (and below) a corridor needing to be subdivided for means of escape purposes (please refer to Figure 4.10)	–	–	O	O
Subdivision of cavities to accord with Table 4.12	–	–	O	O
Edges of cavities and around openings	O	O	O	O

Notes
The symbol O indicates that the provisions apply.
Table 4.11 refers to Table 4.12, which can be checked to verify the maximum dimensions of cavities allowable in non-domestic buildings (purpose groups 2–7).

Table 4.12 Maximum cavity dimensions (purpose groups 2–7)

Cavity location	Surface designations within cavity (excluding pipes, cables, conduits, etc.)	Maximum dimension in any direction (m)
Between roof and ceiling	Any	20
Any other cavity	Class 0 or 1 (Class A1 or A2-s3, d2 or B-s3, d2 or C-s3, d2)	20
	Any other Class	10

Note
The European class is indicated in brackets, although this does not automatically equate with the National class.

- above or below any single room exceeding the dimensions of Table 4.12 where the ceiling cavity above has cavity barriers to the room perimeter and at 40 m intervals;
- above or below an undivided area that exceeds 40 m in either plan dimension subject to the room and cavity being within their own compartment, an automatic fire detection and alarm system being installed in the building and cavity surfaces strictly limited. Reference should also be made to BS 5588: Part 9 (BS 9999, Section 7) if the cavity is to be used as a plenum.

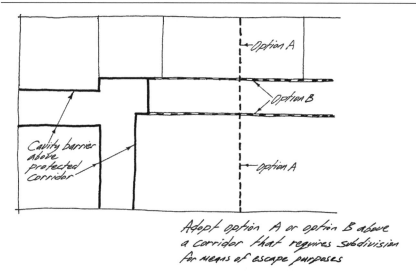

Figure 4.10 Cavity barrier corridor options.

Where an opening needs to be formed in a cavity barrier this should be limited to one of the following:

- minimum 30-minute fire-resisting doors;
- pipes meeting the provisions of Section 10;
- cables or conduits containing cables; and
- openings or ducts (unless fire-resisting) fitted with a suitably installed automatic fire damper.

Section 10: Protection of openings and fire-stopping

Once the lines of fire resistance and allowable openings in them have been established, detailed consideration needs to be given to the fire protection at these locations so as to preserve the integrity of the element and thus avoid fire spread. Note that the test criteria outlined in Appendix A of the Approved Document do not address the issue of smoke spread: for example, a fire damper may offer the required period of fire resistance but allow the passage of smoke, especially in the early stages of fire.

For pipe openings in lines of compartmentation and in cavity barriers three alternatives are given:

- adoption of a proprietary sealing system, e.g. an intumescent collar, for a pipe of any diameter; or
- the restriction of the pipe diameter (see Table 14 of the Approved Document), e.g. 160 mm for cast iron or steel; or

- the sleeving of the pipe (i.e. lead, aluminium, aluminium alloy, PVC or fibre-cement) 1 m either side with a non-combustible material.

For ventilation ducts specific reference should be made to BS 5588: Part 9: 1999 *Code of practice for ventilation and air conditioning ductwork* (BS 9999, Section 7), which offers a series of alternatives so as to preserve compartment integrity:

- fire dampers
- fire resisting enclosures
- fire resisting ductwork.

Fire dampers should not be used for kitchen extract ductwork. Adequate means of fixing the damper and to allow access should also be provided. Fire dampers activated by a fusible link are not suitable for protecting escape routes, and careful means of control is required in buildings involving a sleeping risk.

For flues, ducts containing flues, and appliance ventilation ducts passing through or forming part of a line of compartmentation, the flue wall should achieve at least half the required period of fire resistance for that wall or floor and be of non-combustible construction. See also Approved Document J: Heat-producing appliances.

Suitable fire-stopping provisions should be provided at:

- joints of fire-resisting elements;
- openings around pipes, ducts, conduits and cables; and
- where otherwise specified in the Approved Document.

Fire-stopping needs to take account of thermal movement and unsupported spans exceeding 100 mm. Suitable methods include proprietary sealing systems, intumescent mastics, cement mortar, gypsum-based plaster and glass fibre or crushed rock products. Reference should also be made to *Ensuring best practice for passive fire protection in buildings* and *The Red Book* published by the Association for Specialist Fire Protection (ASFP).

Section 11: Car parks and shopping complexes

Car parks used for cars and other light vehicles are known to have a fire load that is not particularly high, and fire spread, subject to ventilation arrangements, is likely to be limited. In these circumstances a number of provisions are made, as follows:

- The relevant provisions of Requirements B1 and B5 should be applied.
- With the exception of surface finishes, fire doors and attendant's kiosk up to 15 m², and shop mobility facilities, non-combustible construction materials should be used.

- Where the car park does not contain basement levels and is naturally cross-ventilated by permanent openings (5% of the floor area) it may be regarded as **open-sided** with respect to the applicable period of fire resistance (Table 4.9, where certain elements under full load design may provide 15 minute fire resistance with suitable Hp/A factors when tested under European test method).
- If the car park cannot be regarded as open-sided then natural cross-ventilation (2.5% of the floor area) should be provided which may utilize ceiling level vents; *or*
- an independent mechanical ventilation system could be utilized, operating at six air changes per hour for petrol vapour extraction and 10 air changes per hour under a fire condition and designed in two parts, each able to extract 50% of the stated rates and each with its own power supply. Extract outlet locations should be equally split at high and low levels, with the fans and ductwork suitably rated for the high temperatures involved. Reference can also be made to BS 7346: Part 7 and BRE Report 368, which offers an alternative approach.
- See also Approved Document F for normal ventilation of car parks.

For individual shops the guidance contained in the Approved Document, and also in BS 5588: Part 11 (BS 9999, Annex E), can be applied. However, where a **shop becomes part of a complex,** which may include covered malls, common service areas and atria for example, then reference should be made to the more detailed guidance found in BS 5588: Part 10: 1991 *Code of practice for enclosed shopping complexes* (BS 9999, Annex E). This identifies alternative measures and additional compensatory features that are likely to be needed. For information these would include:

- unified management of the complex;
- adequate means of escape;
- smoke control – see BRE Report 386 *Design methodologies for smoke and heat exhaust ventilation* (1999);
- compartmentation;
- sprinkler provisions;
- fire alarm systems;
- access for firefighting;
- isolation of building to relevant boundaries, etc.

On a similar principle to restricting fire spread over internal surfaces, the requirement seeks to control heat release rates over external surfaces. The extent of thermal radiation that could pass through a wall is controlled by limiting the extent of unprotected areas. Roof constructions are controlled to restrict fire spread over their surface and penetration from an external fire source.

REQUIREMENT B4: EXTERNAL FIRE SPREAD

1. The external walls of the building shall adequately resist the spread of fire over the walls and from one building to another, having regard to the height, use and position of the building.
2. The roof of the building shall adequately resist the spread of fire over the roof and from one building to another, having regard to the use and position of the building.

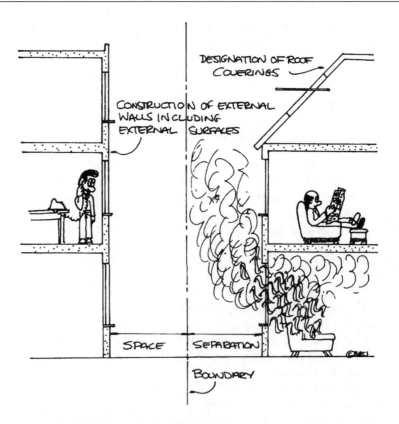

Section 12: Construction of external walls

External walls, as elements of structure, should be able to offer the minimum required period of fire resistance. This is not necessary where the element can be regarded as an **unprotected area,** as defined in Section 13.

The external surfaces of walls should now be considered. A relationship is established for any building, relating boundary isolation distance(s) to the building height and allowable external wall surface classifications as indicated in Table 4.13.

Table 4.13 Provisions for external wall surfaces

Dimension to relevant boundary (m)	Classification for building height of	
	Less than 18 m	18 m or more
Less than 1 1 or more	Class 0 (Class B-s3, d2 or better) No provision, except for purpose group 5 buildings exceeding one storey, which require an Index *I* > 20 (Class C-s3, d2 or better), up to 10 m above ground	Class 0 (Class B-s3, d2 or better) Index *I* > 20 (Class C-s3, d2 or better), up to 18 m, and Class 0 for wall surfaces over 18 m above the ground

Notes

The index (*I*) relates to test specifications in BS 476: Part 6. External insulation (and framework) should be of limited combustibility in a building with a storey above 18 m. For further reference see BRE Report 135 *Fire performance of external thermal insulation for walls of multi-storey buildings* (1988).

The European class is indicated in brackets, although this does not automatically equate with the National class.

An alternative approach could be BRE Report 135 *Fire performance of external thermal insulation for walls of multistorey buildings* (BRE 2003).

Portal frames are widely used in industrial and commercial buildings and can be regarded as acting as single structural elements. In certain circumstances, an external wall close to a boundary for example, these elements of structure may need to offer the required period of fire resistance, which would normally include the rafter section. It has been established that the method of failure under fire loading does not generally warrant the need for fire protection to the rafter section subject to certain provisions, including a satisfactory foundation design. The Approved Document contains no guidance in this regard but makes reference to *P313 Single storey steel framed buildings in fire boundary conditions* (2002), available from the Steel Construction Institute. The SCI recommendations need not be followed if the building is fitted with a sprinkler system. Note that some existing buildings may have been designed with: a rigid column/base fixing; masonry enclosure of column; and some form of roof smoke venting.

Section 13: Space separation

So as to reduce the risk of fire spread across a boundary from one building to another the guidance limits the extent of external wall openings or **unprotected areas** in relation to the isolation distance of the building to the **relevant boundary** (including a **notional boundary**). The first step with any proposal is to establish the location of the relevant boundaries, which are those coinciding with, parallel to or at an angle not exceeding 80° from the building and may be taken to the centreline of a road, railway, canal or river. The space separation of buildings on the same site is not normally regarded as a risk from a building regulation point of view. The two exceptions are:

- if either of the buildings facing each other, new or existing, are in the residential or assembly and recreation purpose groups; or
- if more than one building is constructed on the same site but they are operated/managed by different organizations.

If so, a **notional boundary** needs to be located between the buildings so that both will comply with the provisions for space separation.

Unprotected areas are external wall openings, including windows and doors, external wall panels that do not offer the required period of fire resistance, and external walls with a combustible surface (excluding Class 0 or European class B-s3, d2) more than 1 mm thick (where half the actual area is regarded as unprotected). Certain unprotected areas can be ignored for calculation purposes:

- small areas of 0.1 m^2 and 1 m^2 subject to dimensional restrictions as outlined in Diagram 44 of the Approved Document;
- the external wall of a protected shaft containing a stairway;
- the external wall of a large uncompartmented building where it is more than 30 m above mean ground level; and
- those in an open-sided canopy, minimum 1 m from the relevant boundary.

The Approved Document contains **Methods 1 and 2** to calculate acceptable limits of unprotected area (Tables 4.14 and 4.15) and refers to two other methods, each offering more refined results, which are contained in BRE Report 187 *External fire spread: building separation and boundary distances* (1991).

Table 4.14 Method I – Small residential buildings, purpose groups I and 2

Minimum isolation distance between building and relevant boundary (m)	Maximum unprotected areas (m²)
I	5.6
2	12
3	18
4	24
5	30
6	No limit

Note
Building not to exceed three storeys or 24 m in length.

Table 4.15 Method 2 – Other small buildings or compartments not exceeding 10 m in height

Minimum isolation distance between building and relevant boundary (m)		Maximum unprotected area percentage
Residential, office, assembley and recreation, open-sided car park	Shop and commercial, industrial, storage and other non-residential	
N/A	1	4
1	2	8
2.5	5	20
5	10	40
7.5	15	60
10	20	80
12.5	25	100

Note
The isolation distances may be halved if the building is fitted throughout with an automatic sprinkler system following the guidance in paragraph 0.16 of the Approved Document, subject to a minimum distance of 1 m.

Method 3 – Enclosing rectangles (or geometric method) is contained in Part 1 of BRE Report 187. A plane of reference is chosen and the smallest rectangle (taken from tables in the report) is drawn around the extent of unprotected areas to a particular elevation or compartment. The unprotected areas are expressed as a percentage of the rectangle, from which an isolation distance can then be read from the tables. Reduced isolation distance values for residential, office and assembly and recreation, which are a lower fire risk, are quoted in brackets.

Method 4 – Aggregate notional areas (or protractor method) is also contained in Part 1 of BRE Report 187. This method views the building from a series of points along the boundary and calculates the visible unprotected areas by multiplying the notional areas involved by factors that are dependent on boundary isolation distances. A special 'protractor' is used for this purpose. Aggregate notional area limits are given for residential, office and assembly and recreation uses ($210 \, m^2$) and for shop and commercial, industrial, storage and other non-residential uses ($90 \, m^2$).

Method 5 – Heat radiation and building separation is the final method referred to in Part 2 of BRE Report 187. The method applies first-principle techniques to establish the potential radiation hazard that one building may have on another and to calculate a satisfactory isolation distance to the relevant boundary or between the buildings. This is the most refined method and can give rise to the most accurate isolation distances.

Section 14: Roof coverings

The guidance within this section aims to restrict fire spread across the roof covering of a building, and the penetration of that covering, where it has been exposed to fire from the **outside**. Provisions are made to limit the use of roof coverings dependent on their **designation** (under test to BS 476: Part 3: 2004, first letter indicating penetration and second letter flame spread) and isolation distance to the **relevant boundary** (see Table 16 of the Approved Document). Note that comparative European data/testing are now included in BS EN 13501-5: 2005. For example, no restrictions apply to a roof covering designated AA, AB or AC (B_{ROOF}(t4)), although if a BA, BB or BC (C_{ROOF}(t4)) designated covering is to be used then this should be at least 6 m from the relevant boundary. Note that the wall separating a pair of semi-detached houses is not regarded as a relevant boundary for the purposes of Section 14.

For rooflights incorporated within roof coverings consideration should be given to the following aspects:

- Plastic rooflights with at least a Class 3 lower surface or those formed with TP (a) rigid or TP (b) thermoplastic materials should be spaced with at least 3 m between them and accord with the limitations of use stated in Tables 17 and 18 of the Approved Document, where a minimum 6 m dimension to the relevant boundary is required.
- Rooflights formed with rigid thermoplastic sheet products made from polycarbonate or from unplasticized PVC achieving a Class 1 rating for surface flame spread can be regarded as AA designation.
- Rooflights utilizing unwired glass can be regarded as AA designation if the glass is at least 4 mm thick.
- Thatch and wood shingles can be regarded as AD, BD or CD (E_{ROOF}(t4)).

REQUIREMENT B5: ACCESS AND FACILITIES FOR THE FIRE SERVICE

1. The building shall be designed and constructed so as to provide reasonable facilities to assist firefighters in the protection of life.
2. Provision shall be made within the site of the building to enable fire appliances to gain access to the building.

Until the introduction of Requirement B5 the fire authority would control the provisions necessary to ensure vehicle and personnel access up to and into the building for rescue and firefighting purposes. The building control authority now takes on this role, as far as it applies to the building design, of checking compliance with the Requirement subject to agreed consultation with the fire authority. The degree of access provisions needed to protect the life of firefighting personnel depends mainly on the size of the building. Four sections address the issues controlled under the Requirement. Appendix F addresses the issue of the fire behaviour of insulated core panels used for internal structures.

Section 15: Fire mains and hydrants

The provision of fire mains within the building enables the connection of hoses for water supply to fight fire internally. **Wet** fire mains are permanently charged with water and normally supplied by tanks and pumps located within the building (with an emergency replenishment facility for pumping fire service appliances). **Dry** fire mains are normally empty and supplied by pumping fire service appliances.

Wet or dry fire mains should be provided as follows:

- in firefighting shafts;
- in buildings where vehicle access is not provided in accordance with Table 4.16;
- in buildings with a floor level over 50 m above ground or fire service vehicle access level, where a wet rising main is required; and
- in lower buildings where fire mains are provided, they may be wet or dry.

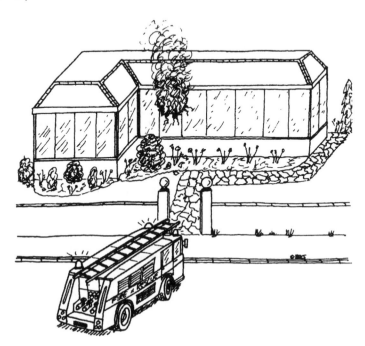

Fire main outlets should be sited within each firefighting lobby between the accommodation and the firefighting shaft. Specific design guidance for fire mains is contained in BS 9990: 2006.

For the first time guidance on the provision of fire hydrants has been incorporated within the Approved Document. Buildings, with compartment sizes over 280 m^2, should be provided with additional private hydrants, clearly marked, where more than 100 m from an existing hydrant:

- buildings with fire mains – within 90 m of dry riser inlet;
- buildings without fire mains – within 90 m of entry point and not more than 90 m apart.

If a piped water supply or sufficient pressure is not available then a charged static water tank or spring/river/canal/pond (with suitable hardstanding for pumped appliance access) should be provided with a minimum capacity of 45 000 l at all times. Any other alternative supply would need to be agreed with the fire authority.

Section 16: Vehicle access

Access to the exterior of the building is required for **pumping appliances** and the deployment of **high-reach appliances** (e.g. turntable ladders and hydraulic platforms) to floor levels over 11 m above the ground level, the extent of which depends on building size. For buildings not fitted with fire mains vehicle access should be provided in accordance with Table 4.16, which relates floor area and height to give the percentage of the building perimeter that requires access.

Table 4.16 Fire appliance access to buildings not fitted with fire mains (excluding blocks of flats)

Total building floor area (m2)	Height of top storey floor level above ground level (m)	Accessible perimeter percentage required
Up to 2000	Up to 11 (pumping appliance)	See note below
	Over 11 (high-reach appliance)	15
2000–8000	Up to 11 (pumping appliance)	15
	Over 11 (high-reach appliance)	50
8000–16000	Up to and over 11	50
16000–24000	Up to and over 11	75
Over 24000	Up to and over 11	100

Note

These small buildings should have vehicle access to within 45 m of any point on the building footprint (including dwelling-houses and flats) or to 15% of the perimeter, whichever is less onerous. The footprint is the maximum aggregate plan perimeter, at any level, excluding walls common with other buildings. Interior access should be via a minimum 750 mm wide door located within the perimeter percentage, which can be used for normal escape purposes.

Table 4.17 Fire appliance access route specification

Minimum requirements	Pumping appliance	High-reach appliance
Road width between kerbs (m)	3.7	3.7
Gateway width (m)	3.1	3.1
Turning circle between kerbs (m)	16.8	26.0
Turning circle between walls (m)	19.2	29.0
Clearance height (m)	3.7	4.0
Carrying capacity (t)	12.5	17.0

Note
Any 'dead-end' access route exceeding 20 m should be provided with a turning circle or hammerhead to accord with the above criteria. For high-reach appliances a clear overhead zone should be maintained in front of the building elevations which require access, maximum dimension 10 m deep (see Diagram 49 of the Approved Document).

For buildings fitted with dry fire mains, access for pumping appliances should be provided to within 18 m of all main inlets, which should be visible from the appliance. Pumping appliances should also be able to get within 18 m of a building fitted with wet fire mains and within sight of the entrance giving to the main and any inlet for emergency replenishment.

The design of access routes and hardstandings, including any manholes, etc., should accord with the specifications contained in Table 4.17.

Section 17: Access to building for firefighting personnel

In high buildings, where ladder access becomes a problem, and certain other building forms, further provisions are necessary to allow effective firefighting operations from within the building. This is generally achieved by the provision of firefighting shafts, which contain firefighting stairs, lobbies and lifts. In the following circumstances firefighting shafts should be provided to a building with:

(a) a floor over 18 m above ground or fire service vehicle access level;
(b) a basement over 10 m below ground or fire service vehicle access level;
(c) a storey, exceeding 900 m², over 7.5 m above ground or fire service vehicle access level, purpose groups 4, 5 and 6 only;
(d) two or more basement storeys, each exceeding 900 m².

The minimum number of firefighting shafts for cases (a) and (c) above, where floor area exceeds 900 m², is two. For buildings fitted with sprinklers provide sufficient firefighting shafts to allow 60 m from any point on the storey to a fire main outlet. If the building is not fitted with sprinklers then any point on the storey should be within 45 m of a fire main outlet located in a protected stair (i.e. not a firefighting shaft) and within 60 m of a fire main outlet in a firefighting shaft. In both cases measured on a route suitable for laying out hoses.

Each firefighting shaft should incorporate a firefighting stairway and lift (except items (c) and (d) above) approached via a firefighting lobby containing a fire main outlet (except at access level). For specific design guidance reference should be made to BS 5588: Part 5 (BS 9999, Section 6). The guidance is modified for blocks of flats due to the high degree of inherent comparmentation. See also Diagram 52 of the Approved Document.

Note that rolling shutters in compartment walls should have the ability to be opened and closed by the fire service, without the use of a ladder.

Section 18: Venting of heat and smoke from basements

Fires within basement levels can cause serious problems for fire service personnel where smoke and heat would tend to exit via stairways. Suitably located smoke outlets connecting with the external air should therefore be provided for fire service use to each basement level or compartment with the exception of any basement not exceeding a floor area of 200 m² and 3 m below ground level.

Natural smoke outlets should be located at high level to induce cross-ventilation and offer a total minimum clear cross-sectional area of 2.5% of floor area served. Areas of special fire hazard should have their own independent vents.

A suitably designed **mechanical smoke extract** system can be used as an alternative, offering a minimum ten air changes per hour, subject to the basement levels also being provided with a sprinkler system. Also refer to BS EN 12101–3: 2002.

Basement car parks can use the guidance in Section 11.

It should be noted that provisions are not made in the Approved Document for smoke venting to ground and upper floor levels. Certain local Acts and British Standards may require such provisions, which normally seek a clear ventilation area of 2.5% of the floor area in question and the clearance of smoke from stairways. Atria, shopping complexes and buildings to which a fire safety engineering approach have been applied may also require the provision of some form of natural or mechanical smoke control measure.

Appendices

The principal contents of the appendices contained within the Approved Document have been touched upon within the body of the text; however, the appendix titles and the following specific items have been highlighted for information.

Appendix A: Performance of materials and structures

Reference to reaction to fire classifications is also included: A1 (the highest), A2, B, C, D, E or F. Subscript s relates to smoke production and d flaming droplets.

Tables A6, A7 and A8, relating to materials, include European classifications.

Appendix B: Fire doors

Table B1 within Appendix B of the Approved Document now makes reference to the minimum fire resistance of a door when tested to the relevant European standard. An additional classification of S_a is used for all doors where restricted smoke leakage at ambient temperatures is needed.

Appendix C: Methods of measurement

- The measurement of cubic capacity would include any roof space over a compartment.
- The height of the top storey of a building is measured from the upper surface of the top floor (excluding rooftop plant areas) down to ground level on the lowest side of the building.
- The height of a building is measured from the mean roof level down to the mean ground level.

Appendix D: Purpose groups

Table D1 is reproduced earlier within the text as Table 4.1.

Appendix E: Definitions

Most of the definitions concerning Requirements B1–B5 have been highlighted and explained within the text. The following are provided for information:

Atrium – a space within a building, not necessarily vertically aligned, passing through one or more structural floors (excluding a shaft used only for stairs, escalators, lifts or services); also refer to BS 5588: Part 7.

Basement storey – a storey with a floor that at some point is more than 1.2 m below the highest level of ground adjacent to the outside walls.

Circulation space – a space (including a protected stairway) mainly used as a means of access between a room and an exit from the building or compartment.

Common stair – an escape stair serving more than one flat or maisonette.

Compartment (fire) – a building or part of a building, comprising one or more rooms, spaces or storeys, constructed to prevent the spread of fire to or from another part of the same building, or an adjoining building.

Direct distance – the shortest distance from any point within the floor area, measured within the external enclosures of the building, to the nearest storey exit ignoring walls, partitions and fittings, other than the enclosing walls/partitions to protected stairways.

Dwelling – a unit of residential accommodation occupied (whether or not as a sole or main residence):

(a) by a single person or by people living together as a family; or
(b) by not more than six residents living together as a single household, including a household where care is provided for residents.

Emergency lighting – lighting provided for use when the supply to the normal lighting fails.

Escape lighting – the part of emergency lighting that is provided to ensure that the escape route is illuminated at all material times.

Escape route – forming part of the means of escape from any point in a building to a final exit.

External wall (or side of a building) – includes a part of a roof pitched at an angle of more than 70° to the horizontal, if that part of the roof adjoins a space within the building to which persons have access (but not access only for repair or maintenance).

Fire door – a door or shutter (including a cover or other form of protection to an opening), provided for the passage of persons, air or objects, which, together with its frame and furniture as installed in a building, is installed (when closed) to resist the passage of fire and/or gaseous products of combustion, and is capable of meeting specified performance criteria to those ends.

Fire-resisting (fire resistance) – the ability of a component or construction of a building to satisfy, for a stated period of time, some or all of the appropriate criteria specified in the relevant part of BS 476.

Fire separating element – a compartment wall, compartment floor, cavity barrier and construction enclosing a protected escape route and/or a place of special fire hazard.

Gallery – a floor or balcony that does not extend across the full extent of a building's footprint and is open to the floor below (e.g. a raised storage area).

Material of limited combustibility – a material performance specification that includes non-combustible materials, and for which the relevant test criteria are set out in Appendix A, paragraph 9 (and Table A7) of the Approved Document.

Means of escape – structural means whereby (in the event of fire) a safe route or routes is or are provided for persons to travel from any point in a building to a place of safety.

Non-combustible material – the highest level of reaction to fire performance. The relevant test criteria are set out in Appendix A, paragraph 8 (and Table A6) of the Approved Document.

Open spatial planning – the internal arrangement of a building in which more than one storey or level is contained in one undivided volume, e.g. split level floors. For the purposes of this document there is a distinction between open spatial planning and an atrium space.

Platform floor – a floor (including an access or raised floor) supported by a structural floor, but with an intervening concealed space that is intended to house services.

Protected corridor/lobby – a corridor or lobby that is adequately protected from fire in adjoining accommodation by fire-resisting construction.

Protected entrance hall/landing – a circulation area consisting of a hall or space in a dwelling, enclosed with fire-resisting construction (other than any part that is an external wall of a building).

Protected stairway – a stair discharging through a final exit to a place of safety (including any exit passageway between the foot of the stair and the final exit) that is adequately enclosed with fire-resisting construction.

Storey – includes:

(a) any gallery in an assembly building, purpose group 5; and
(b) any gallery in any other type of building if its area is more than half that of the space into which it projects; and
(c) a roof, unless it is accessible only for maintenance and repair.

Storey exit – a final exit, or a doorway giving direct access into a protected stairway, firefighting lobby or external escape route.

Travel distance – the actual distance to be travelled by a person from any point within the floor area to the nearest storey exit, having regard to the layout of walls, partitions and fittings (unless otherwise specified, e.g. as in the case of flats).

Appendix F: Fire behaviour of insulating core panels used for internal structures

The introduction of this appendix reflects the recent and extensive research undertaken on the subject. Insulating core panels are now widely used for a range of internal structures. Examples include cold rooms and 'clean' environment enclosures. The panels forming these free-standing/structural enclosures are generally of sandwich construction with facing sheets of galvanized steel. Insulants used include: expanded or extruded polystyrene, polyurethane, mineral fibre, polyisocyanurate and modified phenolic.

Once exposed to heat from a fire, polymeric core materials (not mineral fibre) generate large quantities of smoke. The overall integrity of the system used under fire conditions requires careful consideration. General design guidance is given within Appendix F, where a risk assessment approach is recommended. Specific reference is also made to *Design, construction, specification and fire management of insulated envelopes for temperature controlled environments* published by the International Association of Cold Storage Contractors (European Division).

Examples of core materials for a range of applications are given:

- mineral fibre – cooking areas, hot areas, bakeries, fire breaks, fire-stop panels and general fire protection;
- all cores – chill and cold stores, blast freezers, food factories and clean rooms.

Appendix G: Fire safety information

Under the new Building Regulation 38 fire safety information must be given to the **responsible person** once the building work is complete or at the time of occupation. Although this only applies to the erection, extension or change of use of a **relevant building**, the Building Control Body should establish that this information has been provided for all controlled building work, including material alterations.

The appendix offers guidance on the level of information likely to be required. For **simple buildings** 'as-built' plans incorporating the following should be sufficient:

- escape routes, including fire door/exit locations and door ironmongery;
- lines of fire resistance, including compartmentation, cavity barriers and areas of high fire risk;
- fire alarm, emergency lighting provisions and fire exit signage;
- first aid and other firefighting equipment, including sprinklers/isolating valves, fire hydrant locations and smoke-control provisions;
- specification of fire safety equipment and maintenance required; and
- fire safety management arrangements, including evacuation of disabled persons.

For **complex buildings** a more detailed record of the fire safety strategy would be required. Guidance can also be found in BS 5588–12: 2004 *Fire precautions in the design, construction and use of buildings: managing fire safety* (BS 9999, Section 9). These records should include those listed for simple buildings and:

- the fire safety strategy and any fire risk assessments;
- fire safety management arrangements;

- escape strategy, including routes and assembly points;
- all passive fire safety measures, e.g. compartmentation;
- all active fire safety measures, e.g sprinklers or smoke control provisions; and
- specifications of fire safety equipment to include on-going control/maintenance, commissioning certification and any other details appropriate for the specific building.

Appendix H: Standards and other publications referred to

The majority of publications relevant to Requirements B1–B5 have been listed within the text. European test methods and classifications are now fully listed in Appendix H of the Approved Document.

Chapter 5

Approved Document C: Site preparation and resistance to contaminants and moisture

Requirement C1 seeks to address the risks associated with unsuitable material on the building site, contaminants on or in the ground (covered by the building and any land associated with the building) and groundwater. This is only as far as is necessary to ensure the health and safety of persons in and about buildings. Requirement C2 deals specifically with the resistance of the building to moisture penetration, including the causes listed below, via the floor, walls and roof.

REQUIREMENT C1: PREPARATION OF SITE AND RESISTANCE TO CONTAMINANTS

1. The ground to be covered by the building shall be reasonably free from any material that might damage the building or affect its stability, including vegetable matter, topsoil and pre-existing foundations.
2. Reasonable precautions shall be taken to avoid danger to health and safety caused by contaminants on or in the ground covered, or to be covered by the building and any land associated with the building.
3. Adequate subsoil drainage shall be provided, if it is needed to avoid:

    (a) the passage of ground moisture to the interior of the building;
    (b) damage to the building, including damage through the transport of water-borne contaminants to the foundations of the building.

4. For the purposes of this requirement, 'contaminant' means any substance that is or may become harmful to persons or buildings including substances that are corrosive, explosive, flammable, radioactive or toxic.

REQUIREMENT C2: RESISTANCE TO MOISTURE

The walls, floors and roof of the building shall adequately protect the building and people who use the building from harmful effects caused by:

(a) ground moisture;
(b) precipitation including wind-driven spray;
(c) interstitial and surface condensation; and
(d) spillage of water from or associated with sanitary fittings or fixed appliances.

Section 0 of the Approved Document offers an introduction to the provisions including the range of hazards, associated with the ground, that need to be taken into account, for example, vegetable matter and tree roots, chemical and biological contaminants, and gas generation from biodegradation of organic matter. Physical hazards include underground storage tanks, unwanted foundations, unstable fill or unsuitable hardcore containing sulphates. Naturally occurring hazards include the radioactive gas radon and gases produced by other soils and minerals. To comply with Requirement C2 the range of moisture sources is listed, i.e. from the ground: driving rain or wind-driven spray; surface condensation from the water vapour generated within the building; and interstitial condensation. Spillage of water, from fixed appliances, etc., also now needs to be considered.

Although not a building regulation requirement, guidance on **flood risk** is contained in the Approved Document. This would normally be a Town Planning issue, although the building could be designed to mitigate the effects of flooding where located in a flood-prone area:

- Provision of adequate subsoil drainage (see Section 3) to overcome high groundwater levels.
- Use of non-return valves and anti-flooding devices (see Section 3) to overcome backflow or surcharging of sewers or drains.
- Exclude groundwater with water-resistant construction and allow access for clearing water to under-floor voids (see Section 4).
- The Approved Document also makes reference to *Improving the flood performance of new buildings – Flood resilient construction* 2007 (CLG/Defra/EA)

For land affected by contaminants reference is made to the National Planning Policy Framework 2012 (CLG).

The authorities that may need to be notified are the Environmental Health Department of the local council, or the Environment Agency. Specific guidance on the assessment of land affected by contaminants is set out in Appendix A of the Approved Document.

Section I: Clearance or treatment of unsuitable material

Prior to the construction of any building on a site the guidance places reliance on the production of a **site investigation**. The extent would be dependent on the scale of the building and consists of these well-defined stages:

- **planning stage** (where objectives are set);
- **desk study** (review of historical, geological and environmental information);
- **site reconnaissance or walkover** (identification of actual/potential hazards); and
- **main investigation/reporting** (to include sampling/testing to verify soil parameters).

A small domestic extension, for example, would only require a desk study and consideration of the excavations once started on site. A major redevelopment would involve a site investigation including groundwater levels, underlying geology, physical hazards, soil parameters, contaminants, etc. The main sources of reference are BS EN 1997 and BS 5930: 1999 *Code of practice for site investigations* and a series of BRE Digests.

Consideration should then be given to **unsuitable materials**.

- Sufficient turf and other vegetable matter should be removed from the site of building, normally to a depth of at least 150 mm. This does not apply in the case of general storage buildings, or a building where the health or safety of those employed would not be affected.
- Building services should be of such design as to protect them against damage from tree roots. For example, below-ground drainage could be suitably encased in concrete, or utilize pipes with flexible joints.
- Mature trees on shrinkable clay subsoils can cause significant structural damage to property. Diagram 1 and Table 1 of the Approved Document locate the distribution of shrinkable clays and (sulphate-bearing) subsoils, London and Gault clays having a high to very high volume change potential. Reference should be made to BRE Digests 241, 242 and 298, and NHBC Standards Chapter 4.2.

- The presence of existing foundations, services, buried tanks, etc., which could endanger persons in or about the building and on any land associated with that building, must be considered and appropriate action taken.
- Fill or made-ground encountered on a site will need to be assessed so as to prevent differential settlement; reference should be made to BRE Digest 427 and BRE Report 424.

Section 2: Resistance to contaminants

Building sites may contain a wide range of contaminants as a consequence of the previous use(s) of the site; these are illustrated in Figure 5.1.

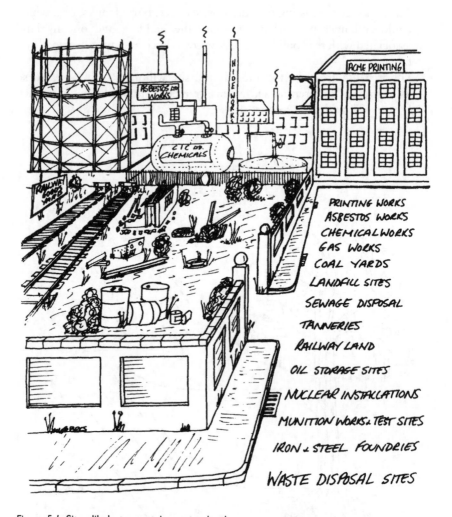

Figure 5.1 Sites likely to contain contamination.

Contaminants may emanate from one of the following sources:

- solids, liquid and gases arising from previous use of land;
- natural contamination by radon and its decay products; or
- landfill gases from buried waste.

The first stage is to identify land that may contain contaminants. This may be possible from local planning records, the refuse authority or local knowledge. Where no records exist, and contaminants in the ground are suspected, reference can be made to Table 5.1, which gives an indication of the signs to look for.

The 2004 edition of the Approved Document makes reference to a substantial amount of guidance on the assessment of contaminated land to show compliance with Requirement C1 (2). To offer a summary of the process of managing land affected by contaminants, Figure 5.2 is taken from the Contaminated Land Research Report CLR 11 – *Model procedures for the management of land contamination* 2004. (Note that this has now been removed from the 2013 edition of the Approved Document).

Solid and liquid contaminants are first discussed within the Approved Document where the concept of the 'source–pathway–receptor' relationship is illustrated. The aim must be to break the pollutant linkages contained in the land with the receptors (i.e. buildings, building materials/services and people). This can be achieved by:

• **treatment** – use of physical, chemical or biological processes to decrease mass, concentration, mobility, flux or toxicity;
• **containment** – blocking by way of covering systems, including vertical barriers to prevent migration;

Table 5.1 Examples of possible contaminants

Signs of possible contaminants	Possible contaminants
Vegetation (absence, poor or unnatural growth)	Metals Metal compounds Organic compounds Gases (landfill or natural source)
Surface materials (unusual colours and contours may indicate wastes and residues)	Metals Metal compounds Oily and tarry wastes Asbestos Other mineral fibres Organic compounds including phenols Combustible material including coal and coke dust Refuse and waste
Fumes and odours (may indicate organic chemicals)	Volatile organic and/or sulphurous compounds from landfill or petrol/solvent spillage Corrosive liquids Faecal animal and vegetable matter (biologically active)
Damage to exposed foundations of existing buildings	Sulphates
Drums and containers (empty or full)	Various

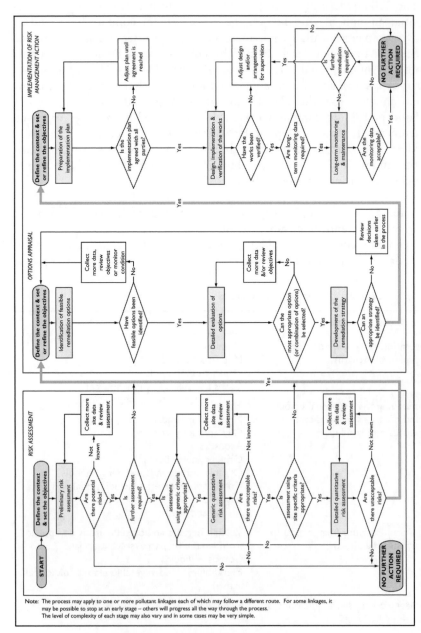

Figure 5.2 The process of managing land affected by contaminants.

- **removal** – excavation and safe removal, target hot spots or a sufficient depth of contaminated material; and
- **protection** – remove receptor by changing the building form or using appropriately designed materials, e.g. suitably sealed imperforate barrier.

With these options in mind a **risk assessment**, undertaken in stages, needs to be carried out. This is fully explained in the Approved Document, the main elements being:

- hazard identification – establishing sources, pathways and receptors;
- hazard assessment – analysing potential for unacceptable risk;
- risk estimation – scale of risk and degree of harm; and
- risk evaluation – are risks acceptable or not.

Supporting guidance can be sourced from BS 5930: 1999; BS 10175: 2011; *Guidelines for combined geoenvironmental and geotechnical investigations* (AGS) and various Environment Agency documents. Remedial measures can then be recommended, where guidance can be obtained from R&D Publication 66 (Environment Agency/NHBC) and a series of CIRIA publications.

Finally, consideration should be given to **aggressive substances** (acids, alkalis, solvents, sulphates and chlorides), **combustible fill** (domestic waste, coal, plastics, etc.), **expansive slags** (blast furnace and steel making) and **floodwater affected by contaminants** (including hydrocarbon permeating into polyethylene pipes).

Methane and other gases from the ground are next discussed and are gases that originate from landfill sites (generally methane and carbon dioxide) or are generated naturally (generally methane, carbon dioxide or nitrogen). Volatile organic compounds (VOCs) can also be present as a result of petrol, oil and solvent spillages. Methane and other gases can migrate through the subsoil and through cracks and fissures into buildings. As mentioned for solid and liquid contaminants a **risk assessment** should be undertaken. Additional investigation may be needed if the building/site associated with the building is:

- within 250 m of a landfill site (or likely to be influenced by it);
- contains wide-scale biodegradable substances (including made-ground/ fill);
- affected by petrol, oil or solvent spillage; and
- subject to naturally occurring methane, carbon dioxide and other hazardous gases (e.g. hydrogen sulphide).

Guidance sources to assist the risk assessment are given, i.e. *Waste Management Paper No. 27* (CIMW); Institute of Petroleum TP 95 1998; BGS Technical Report WP/95/1 1995; CLR 11 and three CIRIA guidance documents.

Remedial measures can now be considered, which may include ongoing site-wide gas control measures. The most likely methods would be **containment, removal** or **protection**. This latter solution would involve a gas-resistant barrier across the building footprint, above a passively ventilated zone. For large commercial buildings expert advice should be sought to ensure adequate ventilation, where mechanical means may be necessary. Guidance is available from BRE/Environment Agency Report BR 414; DETR/Arup Environmental Report; and CIRIA Report 149.

Radon, which is a colourless and odourless radioactive gas, can migrate through subsoil and in turn into buildings, and excessive exposure can increase the risk of lung cancer. The regions susceptible to radon include the Counties of Cornwall, Devon, Somerset, Northamptonshire and Derbyshire and are under regular review. The regions are listed, together with appropriate protective measures, in BRE Report 211 *Radon: guidance on protective measures for new buildings* 2007. Protective measures should also be taken for workplaces where the BRE Report can be taken as a starting point, although reference can be made to BRE Report FB4 *Radon in the workplace* 2011.

Section 3: Subsoil drainage

- Where the site is liable to flooding appropriate steps should be taken. (see notes on Section 0).
- Effective drainage should be provided to protect the building against a high water table (rising to within 0.25 m of the lowest floor level) and the entry of surface water.
- Where an active subsoil drain is cut, it should be sealed if passing under a building, re-routed around the building or rerun to another outfall.

Section 4: Floors next to the ground

This section deals with the need for a floor next to the ground to:

- resist the entry of moisture to the upper surface of the floor, but not in the case of general storage buildings or a building where the health or safety of those employed would not be affected;
- not be damaged as a consequence of moisture from the ground;
- not be damaged by groundwater; and
- resist the passage of ground gases (cross-reference Section 2).

Three types of floor are dealt with, as illustrated in Figure 5.3, together with the risks of interstitial condensation and surface condensation.

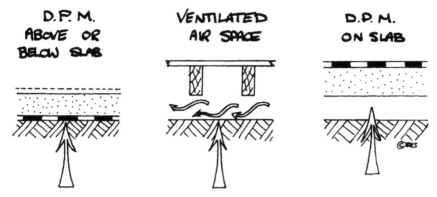

Figure 5.3 Provisions for floors next to the ground.

Floors supported directly on the ground should be constructed to the following guidelines, unless they are subject to water pressure, where the **alternative approach** should be adopted:

• bed of clean hardcore (maximum 600 mm deep), e.g. broken bricks or reject stone, containing no harmful sulphates; and

• minimum 100 mm thick concrete, BS 8500 mix ST2 (or ST4 if reinforcement is being used) or equivalent; and

• damp-proof membrane (DPM) minimum 300 m (1200 gauge) polythene, with joints sealed, lapped with damp-proof courses (DPC) and laid under the slab on a suitable bedding material; or

• the damp-proof membrane laid over the slab, which may be as above or use a three-coat cold-applied bitumen solution protected by the screed or floor finish;

• insulation under slab to resist self-weight and applied loads;

• a timber floor finish directly on a concrete slab should have wood preservative treatment to any timber fixing fillets and be bedded in a material acting as a damp-proof membrane;

• for an **alternative approach** reference can be made to CP 102: 1973 *Protection of buildings against water from the ground,* or BS 8102: 1990 *Code of practice for protection of structures against water from the ground.* (Note that the Approved Document: Basements for dwellings has been withdrawn.)

For a suspended timber floor the following guidelines should be adopted:

• Cover the ground with 100 mm concrete on a hardcore base or with a DPM, as described above, weighted with 50 mm concrete or fine aggregate.

- The ground level under the floor should be above the adjoining ground level or effectively drained.
- The ventilated air space, 1500 mm^2 per metre run or 500 mm^2/m^2 floor area on both sides, should be at least 150 mm to the underside of the floor from the ground cover level and 75 mm to the underside of any wall plates (supporting floor joists) at which location a DPC should be installed.
- Floor boarding to kitchens, bathrooms, utility rooms, etc., where water could be spilt should be moisture resistant, e.g. softwood at least 20 mm thick or suitable chipboard products.
- For an **alternative approach** reference can be made to CP 102: 1973 *Protection of buildings against water from the ground.*

Suspended concrete floors should follow these design guidelines and any reinforcement should be protected from moisture:

- use in-situ concrete, minimum 100 mm thick, 40 mm reinforcement cover, 300 kg/m^3 cement; or
- use precast concrete, with or without infill blocks, reinforcement protected to at least moderate exposure; and
- incorporate a DPM if ground cover level is below adjoining ground level and not effectively drained; and
- provide a ventilated air space (150 mm), 1500 mm^2 per metre run on both sides.

Ground floors and floors exposed from below (resistance to damage from interstitial condensation), i.e. above an open parking space or passageway, will meet the requirement if they are designed and constructed in accordance with:

- Clause 8.5 and Appendix D of BS 5250: 2002; BS EN ISO 13788: 2002; and BRE Report 262 2002.

Floors (resistance to surface condensation and mould growth) will meet the requirement if:

- ground floor is designed/constructed such that the U-value does not exceed 0.7 W/m^2 K at any point; and
- for all floors, the junctions between elements are designed to accord with Accredited Construction Details (downloaded from the planning portal website) or follow the guidance of BRE IP17/01.

Section 5: Walls

Walls should:

- resist the entry of moisture into the building from the ground and from rain and snow (or precipitation) (external walls), but not apply in the case of general storage buildings or a building where the health or safety of those employed would not be affected;
- not be damaged by such entry of ground moisture and rain and snow (external walls); and
- not be adversely affected by interstitial condensation and not promote mould growth or surface condensation.

Three situations are highlighted in the Approved Document guidance, which are illustrated in Figure 5.4.

For internal and external walls that need to resist moisture from the ground a suitable DPC of bituminous material, polyethylene, engineering bricks or slates should be installed, at least 150 mm above adjoining ground level for an external wall, and lapped with any DPM. For a cavity external wall a cavity tray discharging to the outside face may be required, e.g. over a concrete window beam or to the perimeter of a raft type foundation.

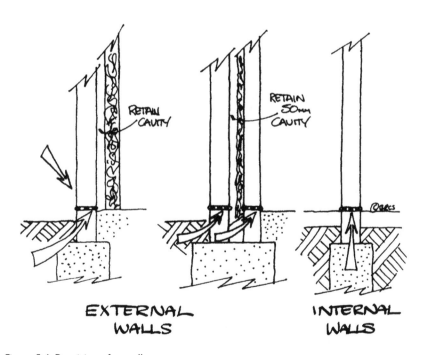

Figure 5.4 Provisions for walls.

An **alternative approach,** where groundwater pressure may present a problem, can be found in BS 8215: 1991 *Code of practice for design and installation of damp-proof courses in masonry construction* or BS 8102: 1990 *Code of practice for protection of structures against water from the ground.*

External walls also need to offer protection from rain and snow: to achieve this one of the following design options may be chosen. For solid walls subject to very severe exposure external cladding should be used, but for those subject to severe exposure or less then adopt the following:

- minimum 328 mm brickwork, 250 mm dense aggregate concrete block-work or 215 mm lightweight aggregate concrete blockwork; and
- minimum 20 mm two-coat rendering, reference BS 5262 1991 and BS EN 998: 2003;
- top of the wall to be protected by a coping and a DPC, where necessary;
- the **alternative approach** is to follow BS 5628: Part 3: 2001.

For cavity external walls:

- the outer leaf should be of brick, block, stone or cast stone; and
- have at least a 50 mm cavity, maintaining a nominal 50 mm residual cavity where partial insulation fill is used, bridged only by wall ties, cavity tray or vertical DPC at openings; and
- an inner leaf of masonry or lining frame, e.g. timber or steel frame;
- the **alternative approach** is to follow BS 5628: Part 3: 2001.

An external wall cavity may be insulated during the course of construction or after the wall is complete. This should be carried out in strict accordance with a current certificate from an appropriate body or a European Technical Approval. Also refer to Driving Rain Map and Table 4 of the Approved Document. The suitability of an existing cavity wall should be assessed in accordance with BS 8208: Part 1: 1985.

For framed external walls the cladding should be separated from the insulation or sheathing by a ventilated and drained cavity with a 'vapour open' membrane located on the inside face of the cavity, able to resist liquid water entry. Care should be taken with the provision of cavity barriers in this form of construction. The possibility of cracking in external masonry walls should be taken into account by reference to BRE Reports 292 and 352, and BS 5628-3: 2001.

Cladding, now included under walls, can take a number of forms, e.g. steel sheeting to a wall or clay tiling to a roof. Both should resist the penetration of rain and snow (precipitation) to the inside of the building and not be damaged by such moisture.

The four cladding design options can be summarized as follows:

- **impervious,** including metal, plastic, glass and bituminous products;
- **weather-resisting,** including natural stone or slate, cement-based products, fired clay and wood;
- **moisture-resisting,** including bituminous and plastic products with joints lapped, if used as a sheet material, and permeable to water vapour unless a ventilated space is provided directly behind the material especially if timber; or
- **jointless materials** and **sealed joints,** allowing for structural and thermal movement.

Note that the suitability of dry joints used in cladding systems will depend on design and wind/rain exposure. In addition, paint and other surface coatings are not regarded as a weather-resisting part of a wall.

The incorporation of insulation should take into account the need for moisture protection and the problems associated with condensation and cold bridging, which are highlighted in the BRE Report 262 2002.

As an **alternative approach** a number of British Standards are referred to:

- BS CP 143: *Code of practice for sheet roof and wall coverings.*
- Part 1: 1958 *for corrugated and troughed aluminium.*
- Part 5: 1964, *zinc*; Part 10: 1973, *galvanized corrugated steel.*
- Part 12: 1970 (1988), *copper*; Part 15: 1973 (1986), *aluminium.*
- Part 16: 1974, *semi-rigid asbestos bitumen sheets.*
- BS 6915: 2001 *for lead.*
- BS 8298: 1994 *for natural stone.*
- BS 8200: 1985 *for walls and steep roofs in general.*
- BS 8219: 2001 *profiled fibre cement.*
- BS 8297: 2000 *non-loadbearing precast concrete cladding.*
- MCRMA Technical Papers 6 and 9.

Additional items of guidance conclude Section 5 and are listed as follows:

- The **joint between walls and door/window frames** should resist the penetration of precipitation to avoid building damage. Suitable damp-proof courses are to be used and for areas in driving rain exposure zone 4 a checked and sealed rebate (minimum 25 mm) should be used in all window and door reveals.
- An unimpeded **door threshold** to meet Part M should have a maximum slope of 15° and an external landing falling away from the door (1:40–1:60). Accessible thresholds in exposed areas should have a drainage channel; see also BRE GBG 47 and *Accessible thresholds in new buildings: guidance for house builders and designers* 1999 (TSO).

- For **external walls (resistance to damage from interstitial condensation)** direct reference is made to BS 5250: 2002 and BS EN ISO 13788: 2002. Particular risks include swimming pools and other buildings where high levels of moisture are generated, where ventilation provisions will be especially important.
- For **external walls (resistance to surface condensation and mould growth)**:
  – design/construct such that the U-value does not exceed 0.7 W/m^2 K at any point; and
  – the junctions between elements are designed to accord with Accredited Construction Details or follow the guidance of BRE IP17/01.

Section 6: Roofs

Roofs are required to resist the same risks as highlighted for walls. The same four design options are also applied for **roofs requiring resistance to moisture from the outside**, as are the **alternative approaches**.

For **roofs (resistance to damage from interstitial condensation)** direct reference is made to BS 5250: 2002; BS EN ISO 13788: 2001; and BRE Report 262 *Thermal insulation: avoiding the risks* 2002. The guidance previously contained in Approved Document F has been removed, although the commentary has been included below for information.

The reason for controlling excessive condensation in roof voids is that it can substantially and permanently reduce the performance of the roof thermal insulation material and the roof structure. Before applying the Approved Document guidance consideration should first be given to the following:

- The ventilation provisions outlined in the Approved Document only relate to **cold roof** constructions, where moisture can permeate the insulation, and not **warm roof** decks, since these should not allow permeation of moisture from within the building.
- Small roofs over porches and bay windows may not require ventilation, owing to their limited risk to health and safety.
- The provisions apply to a roof of any pitch, including a wall at 70° or more.
- Ventilation openings may be in a continuous or intermittent form with a screen, baffle, etc. to prevent the entry of insects, etc.
- Purpose-made components can be used to stop insulation materials restricting air flow paths at eaves and other locations.
- The 2001 edition of BRE Report 262 *Thermal insulation: avoiding risks* contains a wealth of supporting **detailed guidance**, and should be referred to where necessary.

Roofs with a pitch of 15° or more (pitched roofs)

The fundamental principle is to provide adequate cross-ventilation of the roof void, above the insulated ceiling, to remove moisture-laden air. This can be achieved as follows:

- **Pitched roofs** should have eaves-level vents at opposite sides of the roof so as to provide the equivalent of a continuous 10 mm wide opening.
- **Lean-to** (or **mono-pitch**) **roofs** should have the same eaves-level vent and the equivalent of a continuous 5 mm wide opening at high level, e.g. venting tiles or abutment ventilation strip.
- The **alternative approach** is to follow BS 5250: 2002 and BS EN ISO 13788: 2002.

Roofs with a pitch of less than 15° and those where the ceiling follows the pitch of the roof

As with Section 1, cross-ventilation arrangements are needed, which must take into account the reduced roof void area and restriction of air flow paths to sloping ceilings following the pitch of the roof, e.g. a loft room. This can be achieved as follows:

- **Flat roofs** should have eaves-level vents at opposite sides of the roof so as to provide the equivalent of a continuous 25 mm wide opening. A minimum 50 mm clear air space should be retained between the underside of the roof deck and the insulation, utilizing counter-battens where necessary. If the roof plan is not a simple rectangle or the span exceeds 10 m, adopt aggregate ventilation openings of 0.6% of roof plan area.
- **Roofs with ceilings following pitch** should have the same eaves-level vent and a continuous 5 mm wide opening at ridge level, e.g. ridge venting tiles. A minimum 50 mm clear air space should be retained between the underside of the roofing felt and the insulation.
- Where one side of the roof **abuts** an external wall, an abutment ventilation strip, vent cowl(s) or venting tiles may be necessary to preserve cross-ventilation where possible.
- Note that a **vapour check** (or vapour control layer), e.g. foil-backed plasterboard, may help to reduce moisture levels into the roof void, but it should not be regarded as a vapour barrier, and the cross-ventilation provisions should therefore be retained.
- The **alternative approach** is to follow BS 5250: 2002 and BS EN ISO 13788: 2002.

Finally, consideration must be given to **roofs (resistance to surface condensation and mould growth)**:

- design/construct such that the *U*-value does not exceed 0.35 W/m² K at any point; and
- the junctions between elements and the details of openings are designed to accord with the Accredited Construction Details or follow the guidance of BRE IP17/01 or MCRMA Technical Paper 14.

The Approved Document concludes with a substantial list of British Standards and other documents referred to.

Chapter 6

Approved Document D: Toxic substances

This Approved Document contains just one requirement, which deals with the health risk of cavity wall insulating materials that give off formaldehyde fumes.

REQUIREMENT D1: CAVITY INSULATION

If insulating material is inserted into a cavity in a cavity wall reasonable precautions shall be taken to prevent the subsequent permeation of any toxic fumes from that material into any part of the building occupied by people.

The technical solution offered in the Approved Document seeks to minimize, as far as is practicable, the passage of fumes to the occupied parts of the building. A cavity wall may be insulated with urea-formaldehyde foam subject to the following provisions:

- The inner leaf should be built of bricks or blocks.
- The cavity wall should be assessed, before filling, in accordance with BS 8208: *Guide to assessment of suitability of external cavity walls for filling with thermal insulants Part 1: 1985 Existing traditional cavity construction.*
- A current Certificate of Registration of Assessed Capability should be held by the person undertaking the work or under a self-certification scheme.
- The material used should accord with BS 5617: 1985 *Specification for urea-formaldehyde (UF) foam systems suitable for thermal insulation of cavity walls with masonry or concrete inner and outer leaves.*
- Installation should accord with BS 5618: 1985 *Code of practice for thermal insulation of cavity walls (with masonry or concrete inner and outer leaves) by filling with urea-formaldehyde (UF) foam systems.*

Note that careful consideration should be given to the detailing of the cavity wall construction if the foam is to be installed in a recently completed construction. As a combustible insulant reference should also be made to the guidance for Approved Document B, relating to the need for cavity barriers as outlined in Table 4.11.

Chapter 7

Approved Document E: Resistance to the passage of sound

The nuisance of sound is regarded as a health and safety issue for persons living in dwellings and rooms for residential purposes, where the enclosing construction of a dwelling needs to be able to control noise levels emanating from adjoining dwellings, other buildings or other parts of the same building. The occupants of the dwelling should be allowed to follow normal domestic activities, including sleep and rest, without threat to their health from sound sources.

Sound emanates from one of two sources. An **airborne** source, speech or loud speakers for example, vibrates the surrounding air, which in turn sets up vibrations in the enclosing walls and floors. An **impact** source, including footsteps or the movement of furniture, sets up vibrations by direct contact.

Vibrations from both these sources spread via floors, internal walls and the inner leaves of external walls, vibrating the air next to them, which causes new airborne vibrations, which are then heard as sound. The same target for airborne insulation is applied to walls and floors and takes account of low-frequency noise. Figure 7.1 illustrates the principles of direct and flanking sound transmission. (Figure 7.6 provides a summary of sound resistance requirements for dwelling-houses and flats.)

REQUIREMENT E1: PROTECTION AGAINST SOUND FROM OTHER PARTS OF THE BUILDING AND ADJOINING BUILDINGS

Dwelling-houses, flats and rooms for residential purposes shall be designed and constructed in a way that they provide reasonable resistance to the sound from other parts of the same building and from adjoining buildings.

This requirement also includes **rooms for residential purposes**, which are defined as a room, or a suite of rooms, which is not a dwelling-house or a

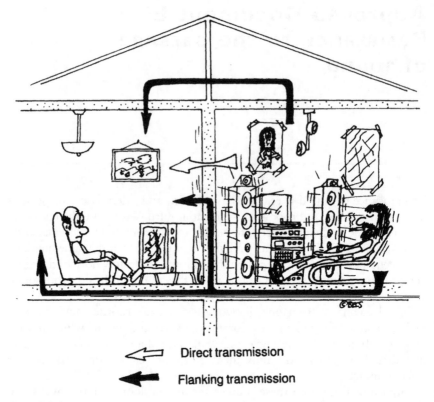

⇐ **Direct transmission**

⬅ **Flanking transmission**

Figure 7.1 Sound transmission paths.

flat and which is used by one or more persons to live and sleep and includes a room in a hostel, a hotel, a boarding house, a hall of residence or a residential home, whether or not the room is separated from or arranged in a cluster group with other rooms, but *does not* include a room in a hospital, or other similar establishment, used for patient accommodation. For the purposes of this definition, a 'cluster' is a group of rooms for residential purposes and is defined as:

(a) separated from the rest of the building in which it is situated by a door that is designed to be locked; and
(b) not designed to be occupied by a single household.

To clarify the application of the provisions of Requirement E1 refer to Figure 7.2.

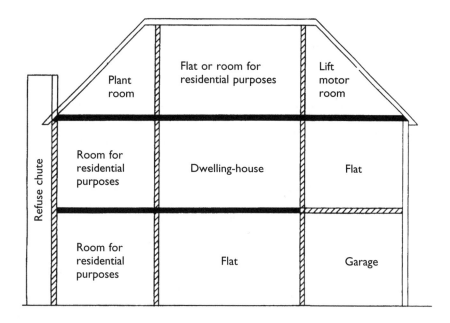

Element to resist airborne sound only

Element to resist airborne and impact sound

Figure 7.2 Walls and floors required to offer sound resistance (Requirement E1).

Note

Regulation 6 applies Requirements E1–E3 to a material change of use where the building is to be used as a dwelling or flat where previously it was not. This also includes a dwelling adjacent to or within another building type.

Sound from external sources is not controlled under Requirements E1–E3. Certain sound levels above those encountered for normal domestic activities may arise: for example, a dwelling located within or adjacent to a night club. In these situations further specialist provisions may be necessary.

REQUIREMENT E2: PROTECTION AGAINST SOUND WITHIN A DWELLING-HOUSE, ETC.

Dwelling-houses, flats and rooms for residential purposes shall be designed and constructed in such a way that:

(a) internal walls between a bedroom or a room containing a water closet, and other rooms; and
(b) internal floors, provide reasonable resistance to sound.

Requirement E2 does not apply to:

- an internal wall that contains a door;
- an internal wall that contains an en suite toilet;
- existing walls and floors in a building that is subject to a change of use.

REQUIREMENT E3: REVERBERATION WITHIN THE COMMON INTERNAL PARTS OF A BUILDING CONTAINING FLATS OR ROOMS FOR RESIDENTIAL PURPOSES

The common internal parts of a building which contain flats or rooms for residential purposes shall be designed and constructed in such a way as to prevent more reverberation sound around the common parts than is reasonable.

Standards for reverberation in the common internal parts of a building containing flats or rooms for residential purposes are set to prevent more reverberation than is reasonable. Only applies to corridors, stairwells, hallways and entrance halls that give access to the flat or room for residential purposes; on site testing is not required. In general, the requirement is met by applying additional absorptive material to the surfaces of the space; refer to the guidance in Section 7.

REQUIREMENT E4: ACOUSTIC CONDITIONS IN SCHOOLS

1. Each room or other space in a school building shall be designed and constructed in such a way that it has the acoustic conditions and the insulation against disturbance by noise appropriate to its intended use.
2. For the purposes of this part 'school' has the same meaning as in Section 4 of the Education Act 1996; and 'school building' means any building forming a school or part of a school.

In simple terms direct reference is made to Section 1 of Building Bulletin 93 *The acoustic design of schools* 2003 (DfES/TSO) to meet Requirement E4.

The performance standards section of the Approved Document clarifies that elements having a separating function, and associated flanking construction, should be built to achieve the sound insulation values stated in Table 7.1. Sound insulation testing may then be necessary as outlined in Section 1. For walls and floors within the dwelling the sound insulation values stated in Table 7.2 should be achieved to comply with Requirement E2.

Table 7.1 Sound insulation values to comply with Requirement E1

	Airborne sound insulation $D_{nT, w} + C_{tr}$ dB (Minimum values)	Impact sound insulation $L'_{nT, w}$ dB (Maximum values)
Purpose-built dwelling-houses and flats		
Walls	45	–
Floors and stairs	45	62
Dwelling-houses and flats formed by material change of use		
Walls	43	–
Floors and stairs	43	64
Purpose-built rooms for residential purposes		
Walls	43	–
Floors and stairs	45	62
Rooms for residential purposes formed by material change of use		
Walls	43	–
Floors and stairs	43	64

Table 7.2 Sound insulation values to comply with Requirement E2

	Airborne sound insulation R_w dB (Minimum values)
Walls	40
Floors	40

Section 1: Pre-completion testing

To demonstrate compliance on site, that the construction methods and details given within Approved Document E have been constructed correctly and provide adequate sound insulation, pre-completion sound testing will need to be carried out to confirm adequate levels of sound resistance have been met. This is by virtue of Regulation 41 Building Regulations 2010 and Regulations 20(1) and (5) Building (Approved Inspectors, etc.) Regulations 2010. Testing should be carried out for:

- purpose-built dwelling-houses and flats, including those formed by material change of use; and
- purpose-built rooms for residential purposes, including those formed by material change of use.

Sound insulation testing should be carried out to the method set out in Annex B of the Approved Document. The person carrying out the building work should arrange for the testing to be carried out, by a test body with appropriate third-party accreditation, preferably UKAS, ANC or European equivalent. The test results must include the following information:

- Address of building.
- Type(s) of property, i.e. dwelling-house, flat or room for residential purposes, and state if the building is a historic building.
- Date(s) of testing.
- Organization carrying out testing, including third-party accreditation number.
- A statement (table format) giving the following information: the rooms used for each test within the set of tests; the measured single-number quantity ($D_{nT,w} + C_{tr}$ for airborne sound insulation and $L'_{nT,w}$ for impact sound insulation) for each test within the set of tests; the sound insulation values that should be achieved; and an entry stating 'Pass' or 'Fail' for each test within the set of tests.
- Brief details of test, including equipment; statement that the test procedures in Annex B have been followed (with any exceptions); source and receiver room volumes; results of tests shown in tabular and graphical forms for third octave bands according to the relevant part of the BS EN ISO 140 series and BS EN ISO 717 series (with (i) single-number quantities and the spectrum adaptation terms, and (ii) D_{nT} and L'_{nT} data from which the single-number quantities are calculated).

The person carrying out the building work should ensure that the guidance on construction given in this Approved Document, or in another suitable source, is followed properly to minimize the chances of a failed test. Where additional guidance is required, specialist advice on the building design should be sought at an early stage.

Criteria for sound testing

Where required, a sound testing regime should be indicated by the Building Control Body to establish on which properties they wish a set(s) of tests to be conducted, on a development. Assuming no tests have failed, the Building Control Body should stipulate at least one set of tests for every ten dwelling-houses, flats or rooms for residential purposes in a group or sub-group (see following text). Where a site comprises only one pair of dwelling-houses, flats or rooms for residential purposes, they should be tested. The Approved Document suggests that, where a development consists of a number of dwellings, one set of tests should be carried out between the first dwelling-houses, flats or rooms for residential purposes scheduled for completion and/or sale in each group or sub-group. This applies regardless of the intended size of the group or sub-group.

It is suggested that testing should be conducted more frequently at the beginning of a series of completions than towards the end, to allow any potential problems to be addressed at an early stage. However, on large developments testing should be carried out over a substantial part of the construction period.

Tests should be carried out once the dwelling-houses, flats or rooms for residential purposes are essentially complete, except for decoration. Impact sound insulation tests should be carried out without a soft covering (e.g. carpet, foam-backed vinyl, etc.) on the floor. Tests should be carried out between rooms or spaces that share a common area of separating wall or separating floor and either side of a separating element to dwelling-houses, flats or rooms for residential purposes.

Tests carried out in rooms for residential purposes should be carried out between the main rooms and measured according to the principles for new buildings and material change of use, but adapting them to suit the circumstances.

Sets of tests

DWELLING-HOUSES (INCLUDING BUNGALOWS)

One set of tests should comprise two individual sound insulation tests (two airborne tests), comprising an airborne sound test between one pair of rooms (where possible suitable for use as living rooms) on opposite sides of the separating wall and a test of insulation against airborne sound between another pair of rooms (where possible suitable for use as bedrooms) on opposite sides of the separating wall.

SETS OF TESTS IN FLATS WITH SEPARATING FLOORS BUT WITHOUT SEPARATING WALLS

One set of tests should comprise four individual sound insulation tests (two airborne tests, two impact tests) for both airborne and impact sound between one pair of rooms (where possible suitable for use as living rooms) on opposite sides of the separating floor. Tests of insulation should be carried out against both airborne and impact sound between another pair of rooms (where possible suitable for use as bedrooms) on opposite sides of the separating floor.

SETS OF TESTS IN FLATS WITH SEPARATING FLOORS AND SEPARATING WALLS

It is preferable that each set of tests contains individual tests in bedrooms and living rooms. Where pairs of rooms on either side of the separating element are different (e.g. a bedroom and a study, a living room and a bedroom), at least one of the rooms in one of the pairs should be a bedroom and at least one of the rooms in the other pair should be a living room.

Grouping and sub-grouping for new buildings

- Dwelling-houses (including bungalows), flats and rooms for residential purposes should be considered as three separate groups.
- Grouping should take account of properties of similar construction, and if there are any significant differences in construction type or configuration then they should be grouped separately, as a sub-group.
- For dwelling-houses (including bungalows), sub-grouping should be by type of separating wall.
- For flats, sub-grouping should be by type of separating floor and type of separating wall.
- Rooms for residential purposes should be grouped using similar principles. If there are any significant differences in the construction of flanking elements (e.g. walls, floors, cavities) and their junctions, then these should be further sub-grouped as necessary.

Sub-grouping for material change of use

The same principles as for new buildings apply, but in practice significant differences are more likely to occur between separating wall and/or separating floor constructions as well as the associated flanking construction(s) in a development. More sub-groups may therefore be required, and group sizes may be smaller. Building control bodies should exercise judgement when setting up sub-groups. In the case of some historic buildings undergoing a material change of use, it may not always be practical to achieve the sound insulation values set out in Tables 7.1 and 7.2. Building control bodies should still be satisfied that everything reasonable has been done to improve the sound insulation. Tests should be carried out, and the results displayed as indicated above.

Action following a failed set of tests

In the event of a failed set of tests, appropriate remedial treatment should be applied to the rooms that failed the test, and remedial action will be taken to justify compliance by the Building Control Body **or** a retest will be carried out. Note that remedial treatment may be needed in rooms other than those in which the tests were conducted.

Alternative to sound testing

Where building work consists of the erection of a dwelling-house or a building containing flats, the requirement for sound testing will not apply to any part of the building should one or more design details be approved by Robust Details Limited (company registration number 04980223) provided the Building Control Body is informed that the construction proposed to be used will be one approved by Robust Details Limited (*LABC not later than the date on which he gives notice of commencement of the work*).

The Building Control Body should be informed:

- of the part or parts of the building in respect of which he is using the design detail;
- of the design detail concerned;
- of the unique number issued by Robust Details Limited in respect of the specified use of that design detail; and
- that the building work carried out in respect of the part or parts of the building identified in the notification is in accordance with the design detail specified in the notification.

To use the **robust details** procedure, as an alternative to testing, follow these steps:

- Register each plot with Robust Details Limited (£30 per plot).
- Give a copy of the Purchase Statement to the Building Control Body with your plans and details, before work commences.
- Build strictly in accordance with the robust details.
- Complete the compliance checklist and make it available to the Building Control Body.

The range of robust details does not initially match the choice provided by the Approved Document guidance. This will, however, expand as manufacturers gain acceptance for their components and systems. For more detailed information relating to registration and the guidance checklists visit www.robustdetails.com or contact by post at: Davy Avenue, Knowhill, Milton Keynes, Bucks, MK5 8NB, Tel: 0870 240 8210.

Section 2: Separating walls and associated flanking constructions for new buildings

Walls need only to resist the passage of airborne sound, and four main options to achieve this are given (Figure 7.3). Included with each wall type are example constructions, which can offer the required level of sound

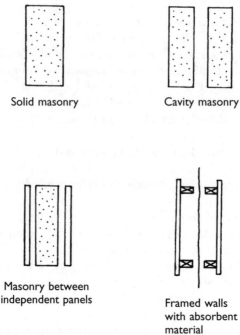

Solid masonry   Cavity masonry

Masonry between
independent panels

Framed walls
with absorbent
material

Figure 7.3 Wall types.
© BRCS

insulation (subject to testing), junction details and a checklist of dos and don'ts that are included in the Approved Document.

Note that a wall separating a habitable room or kitchen from a refuse chute should have a mass per unit area (including any finishes) of at least 1320 kg/m². A wall separating a non-habitable room from a refuse chute should have a mass per unit area (including any finishes) of at least 220 kg/m². Provide good perimeter sealing and minimum mass of 25 kg/m² to corridor doors; a lobby may be preferred to isolate noisy parts of the building.

Wall type 1: Solid masonry

To meet the criteria for solid masonry, the resistance to airborne sound depends mainly on the mass per unit area of the wall.

CONSTRUCTIONS

- **Wall type 1.1** – 215 mm dense aggregate concrete block (minimum block density 1840 kg/m³), laid flat to the full thickness of the wall, 13 mm lightweight plaster (minimum mass per unit area 10 kg/m²) on both room faces, minimum mass per unit area including plaster 415 kg/m².

- **Wall type 1.2** – 190 mm dense aggregate concrete cast *in situ* (minimum concrete density 2200 kg/m³), 13 mm lightweight plaster on both room faces, mass including plaster 415 kg/m².
- **Wall type 1.3** – 215 mm brick (brick density 1610 kg/m³), laid in a bond including headers, 13 mm lightweight plaster on both room faces, mass including plaster 375 kg/m².

JUNCTIONS

- External cavity wall may be of any construction that has an inner leaf of mass per unit area of at least 120 kg/m² unless openings are provided each side of the separating wall, minimum 1 m high and not more than 700 mm from both faces of the separating wall. The cavity wall should be stopped with a flexible closer unless fully filled with mineral wool. The separating wall should be bonded or tied to the inner leaf of the external cavity wall.
- External cavity wall with timber frame inner leaf where the external wall is a cavity wall; the outer leaf of the wall may be of any construction; cavity should be stopped with a flexible closer. Where the inner leaf of an external cavity wall is of framed construction, the framed inner leaf should abut the separating wall, and be tied to it with ties at no more than 300 mm centres vertically. Wall finish of the framed inner leaf of the external wall should be one layer of plasterboard, or two layers where there is a separating floor.
- Seek specialist advice for junctions with an external solid masonry wall.
- Internal masonry walls abutting a type 1 separating wall should have a mass per unit area of at least 120 kg/m² excluding finish.
- Joist hangers should support timber floor joists.
- Concrete floors with a mass per unit area of at least 365 kg/m² may be carried out through a type 1 separating wall; hollow-core concrete plank and concrete block/beam floors should not. Ground bearing floor slabs may continue under a type 1 separating wall.
- Flexible filler (or fire-stop) joint between roof and head of wall (and to head of cavity external wall if applicable).
- Mass of wall within a roof space should be taken full height. If not a habitable room then its mass can be reduced to 150 kg/m², subject to the provision of a sealed ceiling having a mass of 10 kg/m². For lightweight aggregate blocks (density less than 1200 kg/m³) seal one face with cement paint or plaster skim.

Wall type 2: Cavity masonry

The resistance to airborne sound depends on the mass per unit area of the leaves and on the degree of isolation achieved, affected by wall tie and foundation connections and cavity width.

CONSTRUCTIONS

- **Wall type 2.1** – Two leaves of dense 100 mm aggregate concrete (minimum block density 1990 kg/m^3), 50 mm cavity, with 225 mm coursing, 13 mm lightweight plaster (minimum mass per unit area 10 kg/m^2) on both room faces, minimum mass per unit area including plaster 415 kg/m^2.
- **Wall type 2.2** – Two leaves of lightweight 100 mm aggregate concrete (minimum block density 1375 kg/m^3), 75 mm cavity, with 225 mm coursing, 13 mm lightweight plaster (minimum mass per unit area 10 kg/m^2) on both room faces, minimum mass per unit area including plaster 300 kg/m^2.
- **Wall type 2.3** *(only used where minimum 300 mm step/stagger)* – Two leaves of lightweight 100 mm aggregate concrete (minimum block density 1375–1600 kg/m^3), 75 mm cavity, 225 mm coursing, plasterboard (minimum mass per unit area 10 kg/m^2) on both room faces, minimum mass per unit area including plaster 290 kg/m^2.
- **Wall type 2.4** *(only used where minimum 300 mm step/stagger and no separating floor)* – Two leaves of aircrete 100 mm blocks (minimum block density 650 kg/m^3), 75 mm cavity, 225 mm coursing, plasterboard (minimum mass per unit area 10 kg/m^2) or 13 mm plaster on both room faces, minimum mass per unit area including plaster 150 kg/m^2.

JUNCTIONS

- The principles outlined for type 1 walls can also be applied to type 2 walls.
- Concrete floors should generally be built into a type 2 separating wall, but the cavity must not be bridged.
- Mass of wall within a roof space can also be reduced to 150 kg/m^2, although a cavity wall should still be used.

Wall type 3: Masonry between independent panels

The resistance to airborne sound depends partly on the type and mass per unit area of the core construction and partly on the isolation and mass per unit area of the independent panels. The construction consists of either a solid or cavity masonry core wall with independent panels on both sides. These panels and any frame should not be in contact with the core wall.

CONSTRUCTIONS

- **Wall type 3.1** – Solid masonry core of 140 mm dense aggregate concrete block (minimum block density 2200 kg/m^3), 110 mm coursing, with a minimum mass per unit area of core 300 kg/m^2, with independent isolated panels on both room faces of mass per unit area 20 kg/m², i.e. two sheets of plasterboard with joints staggered or composite panel with cellular core (isolate unsupported panels by 35 mm cavity, framed panels by 10 mm cavity).

- **Wall type 3.2** – Solid masonry core of 140 mm dense aggregate concrete block (minimum block density 1400 kg/m³), 225 mm coursing, with a minimum mass per unit area of core 150 kg/m², with independent isolated panels on both room faces of mass per unit area 20 kg/m², i.e. composite panel of two sheets of plasterboard with cellular core.
- **Wall type 3.3** – Cavity masonry core, brickwork or blockwork, 50 mm cavity, with independent isolated panels on both room faces of mass per unit area 20 kg/m², i.e. composite panel of two sheets of plasterboard with cellular core.

JUNCTIONS

- The principles outlined for type 1 and type 2 walls can also be applied to type 3 walls, solid core and cavity walls, respectively.
- Framed internal walls should be fixed via a continuous pad of mineral wool if load bearing or butted up to the independent panels only for non-loadbearing walls. Masonry internal walls should not abut a type 3 separating wall.
- Spaces between the floor joists, supported with hangers, should be sealed with full-depth timber blocking.
- Mass of wall within a roof space can also be reduced to 150 kg/m² and the panels would not be necessary, although any cavity core wall should still be taken full height.

Wall type 4: Framed walls with absorbent material

The Approved Document gives guidance only for a timber framed wall; for steel framed walls advice should be sought from the manufacturer. The resistance to airborne sound depends on the mass per unit area of the leaves, the isolation of the frames and the absorption in the cavity between the frames.

CONSTRUCTIONS

- **Wall type 4.1** – Two frame leaves with absorbent material, minimum 200 mm between inside faces, each lining at least two layers of plasterboard (joints staggered and mass per unit area 10 kg/m²), absorbent material to be unfaced mineral wool batts or quilt (minimum density 10 kg/m³), 25 mm if suspended in the cavity, 50 mm if fixed to one frame and 25 mm if fixed to each frame. Note that any masonry structural core should be connected to only one frame.

JUNCTIONS

- The principles outlined for type 1 walls can also be applied to type 4 walls, subject to the issues below.
- Specialist advice needs to be sought where a type 4 separating wall abuts an external wall with masonry inner leaf, solid masonry wall or an internal concrete floor.
- No restrictions apply to internal framed or masonry walls abutting a type 4 separating wall.
- Block air paths at timber floors with solid timber, continuous ring beam or joists.
- The wall construction should be taken full height and sealed with flexible closer (fire-stop). If not a habitable room the two leaves can be reduced to two layers of plasterboard or a non-rigid cavity closer at ceiling level can be used and a single frame taken up to the roof, with two layers of plasterboard at each side.

Section 3: Separating floors and associated flanking constructions for new buildings

Floors need to resist the passage of airborne sound or both airborne and impact sound. Three main options to achieve this are given (Figure 7.4). As with walls, example constructions (subject to testing) and junction details are given, and a checklist of dos and don'ts are included in the Approved Document.

Each floor type requires one of the ceiling treatments described below, A offering the best sound reduction.

- Ceiling treatment A – Independent ceiling with absorbent material of at least two layers of plasterboard with staggered joints of minimum total mass per unit area of plasterboard 20 kg/m^2 and an absorbent layer of mineral wool (minimum thickness 100 mm, minimum density 10 kg/m^3) laid in the cavity formed above the ceiling.

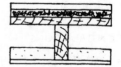

Concrete base with ceiling and soft covering

Concrete base with ceiling and floating floor

Timber base with ceiling and platform floor

Figure 7.4 Floor types.
© BRCS

- **Ceiling treatment B** – Plasterboard on proprietary resilient bars with absorbent material of a single layer of plasterboard, minimum mass per unit area of plasterboard 10 kg/m², fixed using proprietary resilient metal bars. On concrete floors, these resilient metal bars should be fixed to timber battens, with an absorbent layer of mineral wool (minimum density 10 kg/m³) that fills the ceiling void. For fixing details, seek advice from the manufacturer.
- **Ceiling treatment C** – Plasterboard on timber battens or proprietary resilient channels with absorbent material consisting of a single layer of plasterboard, minimum mass per unit area 10 kg/m², fixed using timber battens or proprietary resilient channels. Where resilient channels are used, incorporate an absorbent layer of mineral wool (minimum density 10 kg/m³) that fills the ceiling void.

A soft floor covering is any:

- resilient material, or material with a resilient base, with an overall uncompressed thickness of at least 4.5 mm; or
- floor covering with a weighted reduction in impact sound pressure level (ΔL_w) of not less than 17 dB (measured to BS EN ISO 140–8: 1998 and calculated to BS EN ISO 717–2: 1997).

Floor type 1: Concrete base with ceiling and soft floor covering

The resistance to airborne sound depends mainly on the mass per unit area of the concrete base and partly on the mass per unit area of the ceiling. The soft floor covering reduces impact sound at source.

CONSTRUCTION

- **Floor type 1.1C** – Solid concrete floor slab with a minimum mass per unit area of 365 kg/m² (including shuttering if solid concrete or metal) and including any bonded screed, soft floor covering (essential) and ceiling treatment C or better. Where this construction is to be used in flats with a separating wall type 1 (solid masonry), the separating floor type 1.1C base (excluding any screed) should pass through the separating wall.
- **Floor type 1.2B** – Concrete planks (solid or hollow), with a minimum mass per unit area of planks and any bonded screed of 365 kg/m², soft floor covering (essential), ceiling treatment B or better, all floor joints fully grouted to ensure air tightness. A separating floor type 1.2B base (excluding any screed) should not be continuous through a separating wall type 1 or 3.

JUNCTIONS

- External cavity wall with masonry inner leaf; the outer leaf of the wall may be of any construction, cavity stopped with a flexible closer ensuring adequate drainage (unless fully filled with mineral wool or expanded polystyrene beads), masonry inner leaf of an external cavity wall should have a mass per unit area of at least 120 kg/m² excluding finish, the floor base (excluding any screed) should be built into a cavity masonry external wall and the cavity must not be bridged. These last three issues also relate to internal separating cavity walls.
- External cavity wall with timber frame inner leaf; the outer leaf of the wall may be of any construction, cavity stopped with a flexible closer and two-layer plasterboard inner leaf wall finish.
- Seek specialist advice for junctions with an external solid masonry wall and for junctions with separating wall type 4.
- Where floor type 1.2B is used, and the planks are parallel to the separating wall or an external cavity wall, the first joint should be a minimum of 300 mm from the outer face of the inner leaf. The floor base should not be continuous through a separating wall type 1.
- The floor base should otherwise be continuous (or above) through an internal masonry wall, mass per unit area of at least 120 kg/m² excluding finish, and for type 1 separating walls.
- Pipes and ducts that penetrate a floor separating habitable rooms in different flats should be enclosed for their full height in each flat. Enclosure should be constructed of material having a mass per unit area of at least 15 kg/m², and either line the enclosure or wrap the duct or pipe within the enclosure with 25 mm unfaced mineral fibre. Also consider fire protection/stopping requirements in Part B.

Floor type 2: Concrete base with ceiling and floating floor

The resistance to airborne and impact sound depends on the mass per unit area of the concrete base, as well as the mass per unit area and isolation of the floating layer and the ceiling. The floating floor, which reduces impact sound at source, may consist of a floating layer and a resilient layer as below:

- **Floating floor (A)** – Timber raft of tongued and grooved board (minimum mass per unit area 12 kg/m²), 45 × 45 mm battens, laid loose on the resilient layer (mineral wool with density 36 kg/m³ and minimum thickness 25 mm).
- **Floating floor (B)** – 65 mm sand cement screed (minimum mass per unit area 80 kg/m²), on resilient layer [mineral wool with density 36 kg/m³ and minimum thickness 25 mm *or* one with maximum dynamic

stiffness (measured according to BS EN 29052–1: 1992) of 15 MN/m^3 and minimum thickness of 5 mm under load].

- **Floating floor (C)** – Performance-based approach where generally specialist advice should be sought on the performance and installation of proprietary floating floors.

CONSTRUCTION

- **Floor type 2.1C** – Solid concrete slab with a minimum mass per unit area of 300 kg/m^2 (including shuttering if solid concrete or metal) and including any bonded screed, floating floor A, B or C (essential) and ceiling treatment C or better (essential).
- **Floor type 2.2B** – Concrete planks (solid or hollow), with a minimum mass per unit area of planks and any bonded screed of 300 kg/m^2, regulating screed with all joints fully grouted to ensure air tightness, floating floor A, B or C (essential) and ceiling treatment B or better (essential).

JUNCTIONS

- The principles outlined for type 1 floors can also be applied to type 2 floors, subject to the issues below.
- Pipes and ducts, penetrating a floor separating habitable rooms in different flats, should leave a small gap (approximately 5 mm) between the enclosure and floating layer and seal with sealant or neoprene. Where floating floor A or B is used the enclosure may go down to the floor base, but ensure that the enclosure is isolated from the floating layer.

Floor type 3: Timber frame base with ceiling and platform floor

The resistance to airborne and impact sound depends on the structural floor base and the isolation of the platform floor and the ceiling. The platform floor reduces impact sound at source.

CONSTRUCTION

- **Floor type 3.1A** – Timber structural frame, i.e. joists, with a suitable deck, minimum mass per unit area of 20 kg/m^2, ceiling treatment A (essential) and platform floor, floating layer loose laid on a resilient layer (essential). Floating layer of minimum of two layers of board material, minimum total mass per unit area of 25 kg/m^2, each layer minimum 8 mm thick fixed together (spot bonded with a resilient adhesive or glued/screwed with joints staggered), e.g. 24 mm of cement-bonded particle board or 18 mm T & G timber on 19 mm plasterboard. Resilient layer of 25 mm mineral wool, density of 60–100 kg/m^3.

JUNCTIONS

- External cavity wall with masonry inner leaf; the outer leaf of the wall may be of any construction, cavity stopped with a flexible closer, masonry inner leaf of an external cavity wall should be lined with independent panel (as wall type 3) unless it has a mass per unit area of at least 375 kg/m², any normal method of connecting floor base to wall should be used (block air paths between floor and wall cavities) and ceiling taken through to masonry and sealed. These last three issues also relate to internal separating cavity walls.
- External cavity wall with timber frame inner leaf; the outer leaf of the wall may be of any construction, cavity stopped with a flexible closer, two-layer plasterboard inner leaf wall finish, and any normal method of connecting floor base to wall should be used (seal spaces where joists at right angles to wall).
- Seek specialist advice for junctions with an external solid masonry wall and internal masonry wall.
- Pipes and ducts that penetrate a floor separating habitable rooms in different flats should be enclosed for their full height in each flat. Enclosure should be constructed of material having a mass per unit area of at least 15 kg/m², and either line the enclosure or wrap the duct or pipe within the enclosure with 25 mm unfaced mineral fibre. Also leave a small gap (approximately 5 mm) between the enclosure and floating layer and seal with sealant or neoprene. Also consider fire protection/stopping requirements in Part B.

Section 4: Dwelling-houses and flats formed by material change of use

This section deals specifically with dwellings, houses and flats, which are formed by a material change of use. Guidance for rooms for residential purposes is given in Section 6.

The first consideration is that an existing wall, floor or stair, in a building that is subject to a material change of use, may achieve the performance standards set out in Section 0 without the need for any additional remedial work. This would be the case if the construction was generally similar to one of the constructions described for new dwelling or within 15% of the mass per unit area. In other circumstances it may be possible to use the guidance for new constructions (including flanking constructions) to determine the appropriate remedial treatment, which will result in the construction achieving adequate resistance to sound. Finally, for situations where it is uncertain whether the existing construction achieves adequate resistance to sound, the Approved Document describes one wall treatment, two floor treatments and one stair treatment that can be used to increase sound insulation (Figure 7.5).

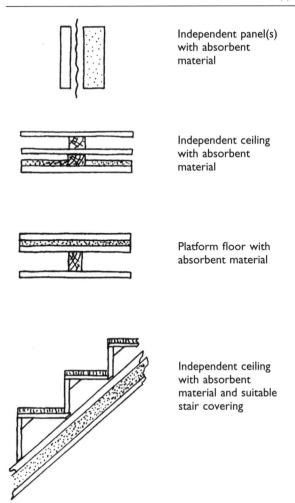

Independent panel(s)
with absorbent
material

Independent ceiling
with absorbent
material

Platform floor with
absorbent material

Independent ceiling
with absorbent
material and suitable
stair covering

Figure 7.5 Treatment for material change of use.
© BRCS

Before any work to existing construction is considered, the following remedial work, prior to any additional treatments, is recommended:

- If the existing floor is timber then gaps in floor boarding should be sealed by overlaying with hardboard or filled with sealant.
- Where floor boards are to be replaced, boarding should have a minimum thickness of 12 mm, and mineral wool (minimum thickness 100 mm, minimum density 10 kg/m^3) should be laid between the joists in the floor cavity.
- If the existing floor is concrete, with an unknown mass or one below 300 kg/m^2, then the mass of the floor should be increased to at least

this figure, any air gaps sealed, and a regulating screed used where necessary.
- If there is an existing lath and plaster ceiling, it can be retained as long as it satisfies Building Regulation Part B: Fire safety.
- Where the existing ceiling is not lath and plaster it should be upgraded as necessary to provide at least two layers of plasterboard with joints staggered, total mass per unit area of 20 kg/m^2.

Extensive remedial work to reduce flanking transmission may also be necessary to achieve adequate resistance to sound. This may involve providing wall linings, and appropriate junction requirements for material change of use, described in the following text.

The separating walls described below may also be used between dwelling-houses or flats formed by material change of use and corridors in order to control flanking transmission and to provide the required sound insulation. However, it is likely that the sound insulation will be reduced by the presence of a door. Any doors should therefore have good perimeter sealing (including threshold where practical) and a minimum mass per unit area of 25 kg/m^2 or a minimum sound reduction index of 29 dB R_w (measured according to BS EN ISO 140–3: 1995 and rated according to BS EN ISO 717–1: 1997). Noisy parts of the building should preferably have a lobby, double door or high-performance doorset to contain the noise. Where this is not possible, nearby flats should have similar protection.

Wall treatment 1: Independent panel(s) with absorbent material

The resistance to airborne sound depends on the form of existing construction, mass of independent panel(s), isolation of the panel(s) and the absorbent material. For existing masonry walls, at least 100 mm thick and plastered on both sides, the independent panel can be used on one side of the existing wall only. In other cases the independent panels should be built on both sides.

CONSTRUCTION

- Minimum mass per unit area of panel (excluding any supporting framework) 20 kg/m^2, consisting of at least two layers of plasterboard with staggered joints, minimum 35 mm gap from masonry core (10 mm gap between any frame, if used, and the face of the existing wall) and minimum 35 mm mineral wool (minimum density 10 kg/m^3) in the cavity.

Floor treatment 1: Independent ceiling with absorbent material

The resistance to airborne and impact sound depends on the combined mass of the existing floor and the independent ceiling, the absorbent material used, the isolation of the independent ceiling and the airtightness of the whole construction.

CONSTRUCTION

- Independent ceiling with minimum mass per unit area of 20 kg/m², consisting of at least two layers of plasterboard with staggered joints, minimum 125 mm gap to underside of existing ceiling (25 mm gap to top of supporting joists) and minimum 100 mm mineral wool (minimum density 10 kg/m³) between the joists. The independent joists should be fixed only to the surrounding walls with additional resilient hangers, attached directly to the existing floor base, if required. Note that because of this, greater depth floor joists may be necessary and hence care needs to be taken to retain sufficient headroom and suitable detailing at window heads (use of pelmet recess).

Floor treatment 2: Platform floor with absorbent material

In this construction resistance to airborne and impact sound depends on the total mass of the floor, the effectiveness of the resilient layer and the absorbent material used.

CONSTRUCTION

- A floating layer is used, minimum two layers of 8 mm board material, minimum total mass per unit area of 25 kg/m², fixed together (e.g. spot bonded or glued/screwed) with joints staggered, laid loose on a resilient layer of mineral wool (minimum thickness 25 mm, density 60–100 kg/m³). Provide 100 mm mineral wool between the joists if used to improve an existing timber floor.

Stair treatment: Stair covering and independent ceiling with absorbent material

Where stairs perform the same separating function as floors they are subject to the same sound requirements. Performance is similar to floors, noting that any cupboard under the stairs should have good airtightness.

CONSTRUCTION

- Use an independent ceiling as described under floor treatment 1, with a soft covering, minimum 6 mm thick, securely fixed over stair treads. If a cupboard is located under the stairs, line the underside within the cupboard with plasterboard (minimum mass per unit area 10 kg/m²) and an absorbent layer of mineral wool (minimum density 10 kg/m^3), including the space above the lining, cupboard walls built with two layers of plasterboard (or equivalent), and use a small, heavy, well-fitted door.

JUNCTIONS

- The principles outlined for previous floor types can also be applied, subject to the issues below.
- Floating floor resilient layer should be turned up at all room edges, with a small gap (approximately 5 mm) between the skirting and the floating layer.
- Where flanking transmission may present a problem, i.e. where external masonry wall has a mass less than 375 kg/m^2, line all adjoining masonry walls with plasterboard or other alternative drylining laminate.

Section 5: Internal walls and floors for new buildings

Four examples of internal wall and floor constructions that meet the laboratory sound insulation values for internal walls and floors for new buildings are given. These have been designed to give insulation against airborne sound; adding a soft covering (e.g. carpet) can help to increase insulation against impact sound. They are listed in order of performance, wall type A generally being the best; all gaps should be filled to avoid air paths between rooms.

CONSTRUCTIONS

- **Internal wall type A** – Timber (75 mm minimum cavity) or metal (45 mm minimum cavity) frames with at least two layers of plasterboard linings on each side (minimum mass per unit area 10 kg/m^2), all joints well sealed.
- **Internal wall type B** – As wall type A, but single-layer plasterboard linings on each side of frame and absorbent material of unfaced mineral wool batts or quilt (minimum thickness 25 mm, minimum density 10 kg/m^3), which may be wire reinforced, suspended in the cavity.
- **Internal wall type C** – Concrete block wall, plaster or plasterboard finish on both sides (minimum mass per unit area, excluding finish, 120 kg/m^2), all joints well sealed.
- **Internal wall type D** – Aircrete block wall, plaster (minimum mass per unit area, including finish, 90 kg/m^2) or plasterboard finish (minimum mass per unit area, including finish, 75 kg/m^2) on both sides, all joints well sealed.
- **Internal floor type A** – Concrete planks (minimum mass per unit area 180 kg/m^2), regulating screed and ceiling finish optional.

- **Internal floor type B** – Concrete beams with infilling blocks (minimum mass per unit area of beams and blocks 220 kg/m^2), bonded screed (minimum 40 mm thickness) and ceiling (treatment C or better from Section 3) required.
- **Internal floor type C** – Timber or metal joist, with wood-based board (minimum mass per unit area 15 kg/m^2), single-layer plasterboard ceiling (minimum mass per unit area 10 kg/m^2), and absorbent material of 100 mm thick mineral wool (minimum density 10 kg/m^3) laid in the cavity.

Notes

- Even if the stair is not enclosed any internal floor should still satisfy Requirement E2.
- Provide good perimeter sealing or a doorset to improve sound insulation at lightweight door locations.
- The layout of rooms should be considered at the early design stage to avoid placing noise-sensitive rooms next to rooms in which noise is generated. Refer to BS 8233: 1999 *Sound insulation and noise reduction for buildings – code of practice.*

Section 6: Rooms for residential purposes

The introduction of rooms for residential purposes has required a separate section to offer example constructions for walls and floors. These should meet the performance standards set out in Table 7.1, and reference is made to constructions in previous sections.

Separating walls in new buildings containing rooms for residential purposes

The following wall types from Section 2 are listed as being suitable:

WALL TYPE 1: SOLID MASONRY

- Wall type 1.1
- Wall type 1.2
- Wall type 1.3.

WALL TYPE 3: MASONRY BETWEEN INDEPENDENT PANELS

- Wall type 3.1
- Wall type 3.2.

Wall types 2 and 4 can be used provided that care is taken to maintain isolation between the leaves. Specialist advice may be needed.

Separating floors in new buildings containing rooms for residential purposes

The most suitable floor types from Section 3 are:

FLOOR TYPE 1: CONCRETE BASE WITH SOFT COVERING

- Floor type 1.1C
- Floor type 1.2B.

Floor types 2 and 3 can be used provided that floating floors and ceilings are not continuous between rooms for residential purposes. Specialist advice may be needed.

Rooms for residential purposes resulting from a material change of use

If the construction was similar (including flanking constructions) to one of those described for solid walls and floors for new constructions in Sections 2 and 3, or within 15% of the mass per unit area of a construction listed in the relevant section, then this would satisfy the requirement.

For situations in which it cannot be shown that the existing construction will achieve the performance standards set out in Table 7.1, Section 4 describes wall, floor and stair treatments to improve the level of sound insulation in dwellings formed by material change of use. These treatments may be used in buildings containing rooms for residential purposes. Specialist advice may also be needed.

Junction details

In general reference should be made to the previous sections to offer guidance for both new buildings and material changes of use. One specific detail noted within the Approved Document guidance is the junction between a solid masonry separating wall of type 1 and the ceiling void/roof space. The wall need not be continuous to the underside of the structural floor or roof subject to: ceiling of at least two layers of plasterboard (minimum total mass per unit area 20 kg/m²); 200 mm minimum layer of mineral wool (minimum density 10 kg/m³) in the roof void; the ceiling is not perforated; and the ceiling joists/plasterboard should not be continuous between rooms for residential purposes.

Section 7: Reverberation in the common internal parts of buildings containing flats or rooms for residential purposes

The common parts of residential developments, which give access to flats and rooms for residential purposes, need to be considered. Section 7 describes how to determine the amount of additional absorption that should be used in corridors, hallways, stairwells and entrance halls. Two methods are set out in the Approved Document.

Method A

- For entrance halls, corridors or hallways, cover an area equal to or greater than the floor area with a Class C absorber or better. For stairwells or a stair enclosure, cover an area equivalent to the combined area of treads, landings (excluding ground floor) and ceiling at the top floor with a Class D absorber, or an area of 50% of this with a Class C absorber. The absorber ratings are determined in accordance with BS EN ISO 11654: 1997. Proprietary acoustic products, especially ceilings, represent a way in which compliance can be achieved.

Method B

- Taking account of the existing absorption provided by all surfaces offers greater flexibility as contained in Method B. This calculation method requires less additional absorption than Method A; the total absorption area, A_T, can be found using the following equation:

$$A_T = \alpha_1 S_1 + \alpha_2 S_2 + \ldots + \alpha_n \Sigma_n$$

where:

S = absorptive material surface area m^2; α= sound absorption coefficient (from Table 7.1 of the Approved Document or BS EN 20354: 1993); and the absorption area A is equal to the product of S and α.

The aim of the calculation is to determine the absorption coefficient, $\alpha_{ceiling}$, needed for the entire ceiling and hence identify a suitable product. A table of generic absorption coefficients for common building materials is given in the Approved Document, together with a worked example and conformation of the reporting methods to be adopted. For a summary of sound resistance Requirement E1, see Figure 7.6.

Section 8: Acoustic conditions in schools

In order to satisfy Requirement E4 for sound insulation, reverberation time and internal ambient noise for schools, the relevant Section 1 of Building

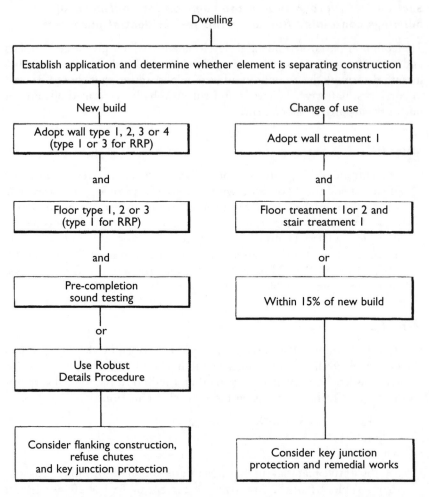

Figure 7.6 Summary of sound resistance Requirement (E1) for dwelling-houses, flats and rooms for residential purposes (RRP).

Bulletin 93 *The acoustic design of schools* (DfES/TSO) should be used. A brief summary is given below.

- Section 1.1 gives the performance standards for new school buildings to comply with the Building Regulations. These provide a good minimum standard for school design. However, on occasions higher standards will be necessary.
- Section 1.2 sets out the preferred means of demonstrating compliance to the Building Control Body.

- Section 1.3 gives the recommended tests to be conducted as part of the building contract.

Although only Section 1 needs to be met to satisfy the requirement, the remaining sections of BB 93 must be considered which briefly are:

- Section 2: Noise control describes how to conduct a site survey and to plan the school to control noise. It also includes recommendations on maximum external noise levels applying to playing fields, recreational areas and areas used for formal and informal outdoor teaching.
- Section 3: Sound insulation gives detailed guidance on constructions to meet the performance standards for sound insulation specified in Section 1.1.
- Section 4: The design of rooms for speech and Section 5: The design of rooms for music give guidance on various aspects of acoustic design relevant to schools.
- Section 7 contains 10 case studies illustrating some of the most important aspects of acoustic design of schools.

Rooms used for nursery and adult/community education within school complexes are covered by Part E, although Part E does not apply to nursery schools that are not part of a school, sixth form colleges that have not been established as schools, and Universities or Colleges of Further and Higher Education. However, many of the acoustic specifications are desirable and can be used as a guide to the design of these buildings. The standards are particularly appropriate for nursery schools as figures are quoted for nursery spaces within primary schools.

Annex A: Method for calculating mass per unit area

Where a mass is specified it is expressed as mass per unit area in kilograms per square metre (kg/m^2). The mass may be obtained from the manufacturer or it may be calculated by the method given in this annex. To calculate the mass per unit area of a masonry leaf use the formula below. This formula is not exact but is sufficient for this purpose.

Formula for calculation of wall leaf mass per unit area

Mass per unit area of a brick/block leaf = mass of co-ordinating area/co-ordinating area

$$= \frac{M_B + \rho_m(Td(L + H - d) + V)}{LH} \quad kg/m^2$$

where:

M_B is brick/block mass (kg) at appropriate moisture content;

ρ_m is density of mortar (kg/m^3) at appropriate moisture content;

T is the brick/block thickness without surface finish (m);

d is mortar thickness (m);

L is co-ordinating length (m);

H is co-ordinating height (m); and

V is volume of any frog/void filled with mortar (m^3).

Annex B: Procedures for sound insulation testing

Describes the sound insulation testing procedure approved by the Secretary of State for the purposes of Regulation 41 of the Building Regulations and Regulations 20(1) and (5) of the Approved Inspectors Regulations.

Annex C: Glossary and Annex D: References

Chapter 8

Approved Document F: Ventilation

Ventilation is the removal of 'stale' indoor air from a building and its replacement with 'fresh' outside air (it is assumed within the Approved Document that the outside air is of reasonable quality). Part F of the Building Regulations is concerned with ensuring that building ventilation systems are provided that will, under normal circumstances, limit the accumulation of moisture which could lead to mould growth and pollutants originating within a building.

A mainly performance-based approach has now been taken to give designers greater freedom in designing systems to meet the performance criteria. The Approved Document has been revised to allow for an increased use of passive stack and mechanical ventilation systems in dwellings, increased levels of thermal insulation, increased air tightness in the construction of buildings and to keep the guidance in step with other design guidance and Regulations. Smoking is not now permitted in most public buildings, and the guidance in the Approved Document does not deal with the products of smoking tobacco.

Greater emphasis is now put onto the **commissioning** of ventilation systems to ensure that performance in use meets the design intentions.

Section 1: Introduction

General guidance on how to use Approved Documents and where to seek further help on Building Regulations matters is set out in Section 1.

Section 2: The Requirements

MEANS OF VENTILATION

REQUIREMENT F1(1):

There shall be adequate means of ventilation provided for people in the building.

Requirement F1 does not apply to a building or space within a building:

(a) into which people do not normally go; or
(b) which is used solely for storage; or
(c) which is a garage used solely in connection with a single dwelling.

REQUIREMENT F1(2):

Fixed systems for mechanical ventilation and any associated controls must be commissioned by testing and adjusting as necessary to secure that the objective referred to in sub paragraph (1) is met.

Section 3: General guidance

This section confirms that Part F does not apply to exempt buildings or extensions as defined in Schedule 2 of the Building Regulations and that certain works can be undertaken under **competent person schemes** in accordance with Schedule 3. Minor works can be carried out without notifying the Building Control Body as listed in Schedule 4 of the Building Regulations, however the work should still comply with the requirements. An example relevant to Part F is replacement parts, or the addition of an output or control device where testing or adjusting is impossible, or would not affect energy efficiency.

For historic and traditional buildings ventilation should be provided as far as reasonably possible without prejudicing the character of the building or increasing the risk of long-term deterioration of its fabric or fittings.

Some key terms used in the Approved Document are reproduced below.

Bathroom – a room containing a bath or shower and, in addition, can also include sanitary accommodation.

Common spaces – those spaces where large numbers of people are expected to gather, such as shopping malls or cinema/theatre foyers. For the purposes of the Approved Document, spaces used principally for circulation (e.g. corridors and lift lobbies in office buildings and blocks of flats) are not common spaces.

Habitable room – a room used for dwelling purposes, which is not solely a kitchen, utility room, bathroom, cellar or sanitary accommodation.

Infiltration – the uncontrolled exchange of air between inside a building and outside through cracks, porosity and other unintentional openings in a building, caused by pressure difference effects of the wind and/or stack effect.

Occupiable room – a room in a building other than a dwelling that is occupied by people, such as an office, workroom, classroom, hotel bedroom, but not a bathroom, sanitary accommodation, utility room or rooms or spaces used solely or principally for circulation, building services plant or storage purposes.

Passive stack ventilation (PSV) – a ventilation device using ducts from terminals in the ceiling of rooms to terminals on the roof that extract air to outside by a combination of the natural stack effect and the pressure effects of wind passing over the roof of the building.

Purge ventilation – manually controlled ventilation of rooms or spaces at a relatively high rate to rapidly dilute pollutants and/or water vapour. Purge ventilation may be provided by natural means (e.g. an openable window) or by mechanical means (e.g. fan).

Ventilation – the supply and removal of air (by natural and/or mechanical means) to and from a space or spaces in a building. It normally comprises a combination of purpose-provided ventilation and infiltration.

Ventilation opening – any means of purpose-provided ventilation (whether it is permanent or closable) that opens directly to external air, such as the openable parts of a window, a louvre or a background ventilator. It also includes any door that opens directly to external air.

Wet room – a room used for domestic activities (such as cooking, clothes washing and bathing) that give rise to significant production of airborne moisture e.g. a kitchen, utility room or bathroom. For the purposes of Part F, sanitary accommodation is also regarded as a wet room.

Whole building ventilation (also called **general ventilation**) – nominally continuous ventilation of rooms or spaces at a relatively low rate to dilute and remove pollutants and water vapour not removed by operation of extract ventilation, purge ventilation or infiltration, *as well as* supplying outdoor air into the building.

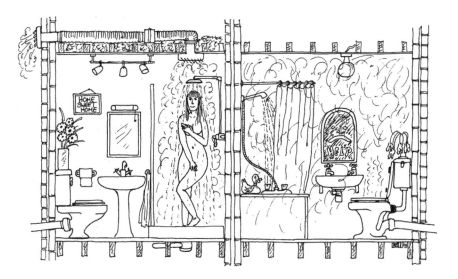

Section 4: Introduction to the provisions

In order to meet the requirement of F1, the ventilation system should:

- extract water vapour from areas where it is produced in significant quantities (e.g. kitchens, utility rooms and bathrooms);
- extract pollutants that are a hazard to health from areas where they are produced in significant quantities (e.g. rooms containing processes or activities that generate harmful contaminants);
- rapidly dilute pollutants and water vapour produced in habitable rooms, occupiable rooms and sanitary accommodation;
- make available over long periods a minimum supply of outdoor air for occupants and disperse residual pollutants and water vapour; such ventilation should minimize draughts and, where necessary, should be reasonably secure and provide protection against rain penetration;
- be designed, installed and commissioned to perform in a way that is not detrimental to the health of the people in the building and be installed to facilitate maintenance where necessary.

Approved Document F adopts the following strategy and gives details of systems that comply with the strategy in Sections 1 and 2 of the Approved Document. Provide:

- **extract ventilation** from rooms where most water vapour and/or pollutants are released, e.g. activities such as cooking or photocopying;
- **whole building ventilation** to provide fresh air to the building and to dilute and disperse residual water vapour and pollutants not dealt with by extract ventilation, as well as removing water vapour and other pollutants that are released throughout the building;
- **purge ventilation** throughout the building to aid removal of high concentrations of pollutants and water vapour released from occasional activities such as painting and decorating or accidental releases such as smoke from burnt food or spillage of water.

This ventilation strategy can be delivered by a natural ventilation system or a mechanical ventilation system or a combination of both. Other ventilation systems may provide acceptable solutions, provided it can be demonstrated to the Building Control Body that they meet the requirement.

Ventilation should be controllable so that it can maintain reasonable indoor air quality and avoid waste of energy; the **controls** can be either manual or automatic.

Manually controlled trickle ventilators should be positioned to avoid discomfort due to cold draughts, typically 1.7 m above floor level. A window with a night latch position is not recommended because of the difficulty of measuring the equivalent area, the greater likelihood of draughts and the potential increased security risk in some locations.

Humidity-controlled devices regulate the humidity of the indoor air to minimize the risk of condensation and mould growth. These are best installed as part of an extract ventilator in moisture-generating rooms, but are not appropriate for sanitary accommodation where the dominant pollutant is normally odour.

Other types of automatic control may be suitable for regulating ventilation devices in dwellings. More sophisticated automatic control systems are available for buildings other than dwellings and may be based on sensors located within the building, e.g. occupancy sensors or indoor carbon dioxide concentration sensors, as an indicator of occupancy level and, therefore, body odour.

Equivalent area has been introduced into the Approved Document instead of **free area** for the sizing of background ventilators because it is a better measure of its **airflow performance.** (The free area of a trickle ventilator is typically approximately 25% greater than its equivalent area.) Ventilators should have the equivalent area (in mm^2 at 1Pa pressure difference) or equivalent area per metre marked on the product in an easily visible location.

Source control, which is the reduction of the release of water vapour and/or air pollutants into the indoor air, is not considered within the Approved Document. Further information is available in BRE Digest 464. House dust mite allergens can trigger allergic reactions in susceptible individuals. Measures for source control are provided in BRE Report BR 417: *Building regulation health and safety.*

In **noisy areas** it may be appropriate to use either sound-attenuating background ventilators or mechanical ventilation solutions, depending on the noise level and any planning conditions. **Noise generated by ventilation fans** can disturb the occupants of the building and so discourage their use, and systems can also create externally emitted noise. The designer should minimize noise by careful design and the specification of quieter products. Guidance is given on how Part F applies to **modular and portable buildings,** which should generally fully comply with Part F applicable at the date of manufacture.

There is new guidance on **air flow rate** testing and commissioning of ventilations systems. The Regulations require all systems capable of being tested and adjusted to be commissioned and a commissioning notice to be given to Building Control. For non-dwellings the commissioning should follow the procedure set out in CIBSE Code M. In dwellings, all mechanical ventilation systems must have the air flow rates measured in accordance with the Domestic Ventilation Compliance Guide and a notice of the results must be given to Building Control. Sufficient information on the correct operation and maintenance of ventilation systems must be given to building owners on completion of new installations.

"…. CLEARLY… THERE'S STILL SOME COMMISSIONING ADJUSTMENTS TO BE DONE..!"

Section 5: New dwellings

The Approved Document shows three main ways of complying with the Requirement:

- provide specified ventilation rates;
- follow the system guidance; or
- use other ventilation systems, provided it can be demonstrated to the Building Control Body that they satisfy the Requirement.

There should be **reasonable access for maintenance**. This should include access for the purpose of changing filters, replacing defective components and cleaning ductwork. The guidance given in Approved Document J should be followed to ensure that, where extract fans or ceiling sweep fans are fitted in buildings containing open flued heat-producing appliances, the appliance burns safely whether or not the fans are operating.

Method 1 – Ventilation rates

The performance will be achieved by providing the airflow rates using:

- extract fans specified in Table 8.1;
- whole building ventilation rates specified in Table 8.2;
- purge ventilation in each habitable room (extract fans can be used for purge ventilation in other rooms, e.g. kitchen and bathrooms, if they are capable of extracting a minimum of four air changes per hour per room directly to outside).

Table 8.1 Extract ventilation rates

Room	Minimum intermittent extract rate (l/s)	Continuous extract Minimum high rate (l/s)	Minimum low rate
Kitchen	30 (adjacent hob) 60 (elsewhere)	13	Total extract rate at least whole building
Utility room	30	8	ventilation rate
Bathroom	15	8	from Table 8.2
Sanitary	6	6	

Table 8.2 Whole building ventilation rates

	Number of bedrooms in dwelling				
	1	2	3	4	5
Whole building Ventilation Rate (l/s)	13	17	21	25	29

Notes
In addition, the minimum ventilation rate should be not less than 0.3 l/s per m² internal floor area (this includes each floor, e.g. for a two-storey building, add the ground and first floor areas).

This is based on two occupants in the main bedroom and a single occupant in all other bedrooms. This should be used as the default value. If a greater level of occupancy is expected, then add 4 l/s per occupant.

Method 2 – Following system guidance

VENTILATION SYSTEMS FOR DWELLINGS WITHOUT BASEMENTS

The performance required for dwellings without basements is achieved by selecting one of four ventilation systems summarized in Table 8.3.

Table 8.3 Overview of system guidance

	Purge ventilation	Mechanical ventilation	Background ventilation	Passive stack ventilation
System 1	•	Intermittent to wet rooms	•	
System 2	•		•	Install to wet rooms
System 3	•	Continuous extract	•	
System 4	•	Continuous supply and extract		

Note
• Provision required.

SYSTEM I – BACKGROUND VENTILATORS AND INTERMITTENT
EXTRACT FANS

- Background ventilators
  For dwellings with more than one exposed façade: for multi-storey
  dwellings, and single-storey dwellings more than four storeys above
  ground level, the **total equivalent area** for the dwelling is given in
  Table 8.4 below; for single-storey dwellings, up to four storeys above
  ground level, take the total equivalent area for the dwelling from the
  table below and add 10000 mm².

  For a dwelling with only a single exposed façade, background ventilators should be located at both high and low positions in the façade
  to provide single-sided ventilation. The total equivalent area at a high
  position is given in the main table. In addition, the same total equivalent
  ventilator area should be repeated and located at least 1 m below the
  high ventilators.
- Intermittent extract
  Note that a continuously running single room heat recovery ventilator
  could be used in a wet room instead of an intermittent fan. It should
  use the minimum high rate in Table 8.1 and 50% of this value as the
  minimum low rate. No background ventilator is required in the same
  room as a single room heat recovery ventilator. The total equivalent
  background ventilator area can be reduced by 2500 mm² for each room
  containing a single room heat recovery ventilator.

Table 8.4 Equivalent ventilator area for dwellings (mm²) for a dwelling with any design
air permeability

Total floor area (m²)	Number of bedrooms				
	1	2	3	4	5
<50	35 000	40 000	50 000	60 000	65 000
51–60	35 000	40 000	50 000	60 000	65 000
61–70	45 000	45 000	50 000	60 000	65 000
71–80	50 000	50 000	50 000	60 000	65 000
81–90	55 000	60 000	60 000	60 000	65 000
91–100	65 000	65 000	65 000	65 000	65 000
>100	Add 7000 mm² for every additional 10 m² floor area				

Notes
The equivalent area of a background ventilator should be determined at a 1 Pa pressure difference,
using the appropriate test method.
    This is based on two occupants in the main bedroom and a single occupant in all other bedrooms.
For a greater level of occupancy, assume greater number of bedrooms (i.e. assume an extra bedroom
per additional person). For more than five bedrooms, add an additional 10 000 mm² per bedroom.

Table 8.5 Equivalent ventilator area for dwellings (mm²) for a dwelling with a design air permeability leakier than 5 m³/hr/m² at 50Pa

Total floor area (m²)	Number of bedrooms				
	1	2	3	4	5
<50	25 000	35 000	45 000	45 000	55 000
51–60	25 000	30 000	40 000	45 000	55 000
61–70	30 000	30 000	30 000	45 000	55 000
71–80	35 000	35 000	35 000	45 000	55 000
81–90	40 000	40 000	40 000	45 000	55 000
91–100	45 000	45 000	45 000	45 000	55 000
>100	Add 5000 mm² for every additional 10 m² floor area				

Notes

The equivalent area of a background ventilator should be determined at a 1 Pa pressure difference, using the appropriate test method.

This is based on two occupants in the main bedroom and a single occupant in all other bedrooms. For a greater level of occupancy, assume greater number of bedrooms (i.e. assume an extra bedroom per additional person). For more than five bedrooms, add an additional 10 000 mm² per bedroom.

SYSTEM 2 – PASSIVE STACK VENTILATION (PSV)

- Provide 125 mm diameter PSV's with an internal cross sectional area of 12000 mm² to kitchens, utility rooms, bathrooms and sanitary accommodation.
- Background ventilators

  The equivalent area of background ventilators should be taken from Tables 8.4 or 8.5 depending on the design air permeability.

  The minimum equivalent area must always be at least equal to the total maximum cross-sectional area of all the PSV ducts. For a dwelling with only a single exposed façade, the dwelling should be designed such that the habitable rooms are on the exposed façade.

SYSTEM 3 – CONTINUOUS MECHANICAL EXTRACT

- Calculate the whole building ventilation rate from Table 8.2 and the whole dwelling air extract rate at maximum operation by summing the individual room rates for minimum high rate from Table 8.1. The required extract rates are as follows: The maximum rate (e.g. 'boost') should be at least the greater of the whole building ventilation rate and the whole dwelling air extract rate. Note that the maximum individual room extract rates should be at least those given in Table 8.1 for minimum high rate, and the minimum rate should be at least the whole building ventilation rate.
- Background ventilators
The need for background ventilators will depend on the air permeability of the dwelling, it is recommended that controllable background ventilators having a minimum equivalent area of 2500 mm^2 are fitted in each room, except wet rooms from which air is extracted.

SYSTEM 4 – CONTINUOUS MECHANICAL SUPPLY AND EXTRACT WITH HEAT RECOVERY (MVHR)

- For any design air permeability determine the whole building ventilation rate from Table 8.2. As an alternative, where the design air permeability is less than 5 m^3/hr/m^2 at 50 Pa allow for infiltration by subtracting from this value:

  - 0.04 l (5 m^3) at 50 Pa × gross internal volume of the dwelling heated space (m^3).

- Calculate the whole dwelling air extract rate at maximum operation by summing the individual room rates for minimum high rate from Table 8.1.

  The maximum extract rate (e.g. 'boost') should be at least the greater of the above figures. The maximum individual room extract rates should be at least those given in Table 8.1 for minimum high rate, and the minimum air supply rate should be at least the whole building ventilation rate.

For all four systems the guidance on **purge ventilation** is as follows:
  For each habitable room with:

- external walls – reference should be made to Appendix B of the document, which sets out window or external door sizing; or
- no external walls – follow the guidance given on ventilating through an adjoining habitable room or conservatory.

For each wet room with:

- external walls – install an openable window (no minimum size); or
- no external walls – the normal extract provisions will suffice, but the extractor fan should have a 15 minute over-run for system 1 solutions.

Also, for all four systems, purge ventilation can be used in sanitary accommodation instead of an intermittent fan, PSV, etc., provided that the calculations for the required ventilation are adjusted appropriately.

Consideration should also be given to the **location of ventilation devices** in rooms:

- Cooker hoods should be 650–750 mm above the hob surface (or follow manufacturer's instructions).
- Mechanical/PSV extract terminals and extract fans should be placed as high as practicable and preferably less than 400 mm below the ceiling.
- Mechanical supply terminals should be located and directed to avoid draughts.
- Where ducts, PSVs, etc. are provided in a dwelling with a protected stairway, precautions may be necessary to avoid the possibility of the system allowing smoke or fire to spread into the stairway.
- For each system other than PSV, extract fans or terminals should be located in each wet room.
- For system 4 air should normally be supplied to each habitable room. The total supply airflow should usually be distributed in proportion to the habitable room volumes. Recirculation by the system of moist air from the wet rooms to the habitable rooms should be avoided.

Background ventilators, minimum 5000 mm^2 equivalent area, should be located in all rooms for systems 1 and 2 other than wet rooms with a PSV, and in each habitable room for system 3; not required for system 4.

The location of **purge ventilation** is not critical. For the **air transfer between rooms** there should be an undercut of minimum area 7600 mm^2 in all internal doors above the floor finish (equivalent to an undercut of 10 mm for a standard 760 mm width door).

Guidance is then given on **controls for ventilation devices.** For mechanical intermittent extract and mechanical continuous supply or extract/passive stack ventilation:

- Ventilation can be operated manually and/or automatically by a sensor (e.g. humidity sensor, occupancy/usage sensor, detection of moisture/pollutant release). Humidity controls should not be used for sanitary accommodation as odour is the main pollutant. In kitchens, any automatic control must provide sufficient flow during cooking with fossil fuels (e.g. gas) to avoid the build-up of combustion products.

- Any automatic control must provide manual override to allow the occupant to turn the extract on or select maximum 'boost' rate.
- Ensure the system always provides the minimum whole building ventilation provision as specified in Table 8.2.

Background ventilators can be either manually adjustable or automatically controlled, with manual operation for purge ventilation. Manual controls should be within reasonable reach of occupants, i.e. use of pull cords, operating rods or similar devices. Reference should be made to Table 1.6 of the Approved Document for details of the **performance test methods** for each type of ventilator.

VENTILATION SYSTEMS FOR BASEMENTS

For a dwelling that includes a basement, which is connected to the rest of the dwelling above ground by a large permanent opening (e.g. an open stairway), the whole dwelling including the basement should be ventilated using one of the four systems given for dwellings without basements and treated as a multi-storey dwelling. If the basement has only a single exposed façade, while the rest of the dwelling above ground has more than one exposed facade, ventilation systems 3 and 4 are preferred. If systems 1 or 2 are to be used, expert advice should be sought.

For a dwelling that includes a basement which is not connected to the rest of the dwelling above ground by a large permanent opening:

- the part of the dwelling above ground should be considered separately and ventilated in accordance with the above guidance; if the part of the dwelling above ground has no bedrooms, assume it has one bedroom for the purpose of determining ventilation provisions; and
- the basement should be treated separately as a single-storey dwelling above ground, in accordance with one of the four systems given for dwellings without basements; if the basement has no bedrooms, assume it has one bedroom for the purpose of determining ventilation provisions.

A dwelling that comprises only a basement should be treated as a single-storey dwelling above ground following one of the four systems given for dwellings without basements.

VENTILATION OF HABITABLE ROOMS THROUGH ANOTHER ROOM OR A CONSERVATORY

In a habitable room not containing openable windows the Requirement will be met if the room is either ventilated through another habitable room or through a conservatory and incorporates the measures shown in Figure 8.1.

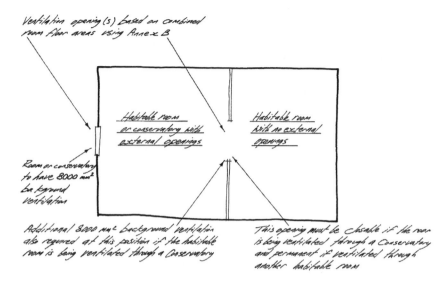

Figure 8.1 Ventilation of habitable rooms through another room or conservatory.

Section 6: Buildings other than dwellings

The Approved Document sets out detailed guidance for offices and car parks and provides a table of other guidance to refer to for other building uses.

The ventilation provisions do not necessarily meet cooling needs. Guidance on the **control of overheating** is provided in Approved Document L2A: New buildings other than dwellings. Where **fresh air supplies** are provided there is a risk of contaminants injurious to health entering the building, so guidance on the siting of air inlets is provided in Appendix F of the Approved Document.

Reference is made to the fact that guidance on design measures to avoid *legionella contamination,* including design features not related to the ventilation of the building, is covered by the Health and Safety Executive (HSE) in *Legionnaires' disease: the control of legionella bacteria in water systems.* Further guidance may be found in CIBSE TM13: *Minimising the risk of Legionnaires' disease.* Also, guidance on recirculated air in air conditioning and mechanical ventilation systems is given by the HSE in Workplace (Health, Safety and Welfare) Regulations 1992 *Approved Code of Practice and Guidance* L24.

Access for maintenance:

- Access should be provided for the replacing of filters, fans and coils and access points for cleaning ductwork should be fitted.

- In a central plant room allow 600 mm space where access is required between plant and 1100 mm where space for routine cleaning is required. Also refer to the Defence Works Functional Standard, Design and Maintenance Guide 08: *Space requirements for plant access operation and maintenance.* Further guidance for the cleaning of ducts is provided by CIBSE *Ventilation hygiene toolkit.*

Offices

To comply, one of the following approaches should be adopted:

- Ventilation should be provided that meets given airflow rates.
- Follow the system guidance in the Approved Document.
- Use the alternative approaches set out in CIBSE guidance.
- Use other ventilation systems provided it can be demonstrated to the Building Control Body that they satisfy the Requirement, e.g. by showing that they meet the moisture and air quality criteria set out in Appendix A of the Document.

VENTILATION RATES

Figure 8.2 illustrates the ventilation rates required for offices.

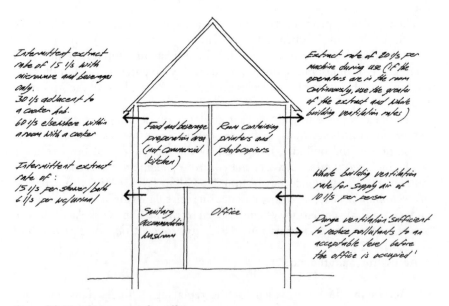

Figure 8.2 Ventilation rates for offices.

NATURAL VENTILATION OF ROOMS

System guidance on achieving the required airflow rates is given in CIBSE Application Manual AM10: 2005 *Natural ventilation in non-domestic buildings*.

MECHANICAL VENTILATION OF ROOMS

The Requirement will be satisfied if the installed performance of the system provides the required airflow rates with the ventilators installed and controlled following the guidance in Table 2.2 of the Approved Document.

Ventilation of other buildings and spaces

Table 8.6 below lists the guidance that should be followed to satisfy the ventilation Requirement for other building uses. Reference should also be made to the Workplace (Health, Safety, Welfare) Regulations 1992.

Ventilation of car parks

The Requirement will be satisfied for car parks below ground level, enclosed-type car parks and multi-storey car parks if the mean predicted pollutant levels are calculated, the ventilation rate designed, and equipment installed, to limit the concentration of carbon monoxide to not more than 30 parts per million averaged over an eight-hour period, and peak concentrations such as by ramps and exits, not more than 90 parts per million for periods not exceeding 15 minutes.

ALTERNATIVE APPROACHES FOR VENTILATION OF CAR PARKS

For **naturally ventilated car parks** well-distributed permanent natural ventilation, at each car parking level, with an aggregate equivalent area equal to at least 1/20th of the floor area at that level, of which at least 25% should be on each of two opposing walls.

For **mechanically ventilated car parks** both permanent natural ventilation openings of equivalent area not less than 1/40th of the floor area and a mechanical ventilation system capable of at least three air changes per hour should be provided, or for basement car parks, the provision of a mechanical ventilation system capable of at least six air changes per hour. In addition, for exits and ramps, where cars queue inside the building with engines running, provisions should be made to ensure a local ventilation rate of at least 10 air changes per hour.

Further guidance can be found in *Code of practice for ground floor, multi-storey and underground car parks* published by the Association for

Table 8.6 Ventilation of other buildings and spaces

Building/space/activity	Regulations and guidance
Animal husbandry	The Welfare of Farm Animals (England) Regulations SI 2000 No.1870 London: The Stationery Office 2000 The Welfare of Farm Animals (England) (Amendment) Regulations SI 2002 No.1646 The Welfare of Farm Animals (England) (Amendment) Regulations SI 2003 No.299 BS 5502 Buildings and Structures for Agriculture See also CIBSE Guide B2: 2005 2.3.24.1 and CIBSE AM10: 2005
Assembly halls	CIBSE Guide B2: 2005 2.3.3; CIBSE AM10: 2005; CIBSE AM13: 2000
Atria	CIBSE Guide B2: 2005 2.3.4; CIBSE AM10: 2005; CIBSE AM13: 2000
Broadcasting studios	CIBSE Guide B2: 2005 2.3.5
Building services plant rooms	Provision for emergency ventilation to control dispersal of contaminating gas releases (e.g. refrigerant leak) is given in paragraphs 23–25 of HSE Guidance Note HSG 202 *General ventilation in the workplace – guidance for employers*. Other guidance in BS 4434:1989 *Specification for safety aspects in the design, construction and installation of refrigeration appliances and systems*
Call centres	CIBSE Guide B2: 2005 2.3.24.2; CIBSE AM10: 2005; CIBSE AM13: 2000
Catering (incl. commercial kitchens)	HSE Catering Information Sheet No. 10, 2000: *Ventilation of kitchens in catering establishments* HSE Information Sheet No. 11, 2000: *The main health and safety law applicable to catering* See also CIBSE Guide B2: 2005 2.3.6 and HVCA DW/172 Specification kitchen ventilation systems
Cleanrooms	CIBSE Guide B2: 2005 2.3.7
Common spaces	These provisions apply to common spaces where large numbers of people are expected to gather, such as shopping malls and foyers. It does not apply to common spaces used solely or principally for circulation. Either: (a) natural ventilation by appropriately located ventilation opening(s)with a total opening area of at least 1/50th of the floor area of the common space; or (b) mechanical ventilation installed to provide a supply of fresh air of 1 l/s per m^2 of floor area
Communal residential buildings	Energy Efficiency Best Practice in Housing, Good Practice Guide GPG 192: *Designing energy efficient multi-residential buildings* See also CIBSE Guide B2: 2005 2.3.8; CIBSE AM10: 2005; CIBSE AM13: 2000
Computer rooms	CIBSE B2: 2005 2.3.9; CIBSE AM10: 2005; CIBSE AM13: 2000
Darkrooms (photographic)	CIBSE Guide B2: 2005 2.3.4.4; CIBSE AM10: 2005; CIBSE AM13: 2000
Dealing rooms	CIBSE Guide B2: 2005 2.3.24.5; CIBSE AM10: 2005; CIBSE AM13: 2000

Factories and warehouses	Factories Act Health and Safety at Work, etc. Act. See also CIBSE B2: 2005 2.3.11; CIBSE AM10: 2005; CIBSE AM13: 2000 Requirements are often exceeded by other criteria such as the ventilation requirements of the particular manufacturing process.
High-rise (non-domestic buildings)	CIBSE Guide B2: 2005 2.3.12; CIBSE AM10: 2005; CIBSE AM13: 2000
Horticulture	CIBSE Guide B2: 2005 2.3.24.6; CIBSE AM10: 2005; CIBSE AM13: 2000
Hospitals and healthcare buildings	NHS Activity database Health Technical Memorandum (HTM) 03 Health Building Notes (HBN) – various CIBSE B2: 2005 2.3.13; CIBSE AM10: 2005; CIBSE AM13: 2000
Hotels	CIBSE Guide B2: 2005 2.3.14; CIBSE AM10: 2005; CIBSE AM13: 2000
Industrial ventilation	*Industrial Ventilation,* 24th Edition, *Manual of Recommended Practice,* American Conference of Government Industrial Hygienists HS(G)37 An introduction to local exhaust ventilation HS(G)54 Maintenance, examination and testing of local exhaust ventilation HS(G)193 COSHH Essentials
Laboratories	CIBSE Guide B2: 2005 2.3.16
Museums, libraries and art galleries	BS 5454:2000; CIBSE Guide B: 2005 2.3.17; CIBSE AM10: 2005; CIBSE AM13: 2000
Plant rooms	CIBSE Guide B: 2005 2.3.18
Prison cells	Refer to National Offender Management Service (NOMS). Home Office, NOMS Property, Technical Services, Room 401, Abell House, John Islip St, London SW1P 4LH
Schools and educational buildings	Ventilation provisions in schools can be made in accordance with the guidance in DfES Building Bulletin 101, *Ventilation of School Buildings* (see www.teachernet.gov.uk/iaq) and in the Education (School Premises) Regulations. Building Bulletin 101 can also be used as a guide to the ventilation required in other educational buildings such as further education establishments where the accommodation is similar to that found in schools, e.g. sixth form accommodation. However, the standards may not be appropriate for particular areas where more hazardous activities take place than are normally found in schools, e.g. some practical and vocational activities requiring containment or fume extraction. The Building Bulletin can also be used for children's centres and other early years settings, including day nurseries, playgroups, etc.
Shops and retail premises	CIBSE Guide B2: 2005 2.3.20; CIBSE AM10: 2005; CIBSE AM13: 2000
Sports centres (incl. swimming pools)	CIBSE Guide B: 2005 2.3.21
Standards rooms	CIBSE Guide B: 2005 2.3.24.7
Sanitary accommodation	As for offices
Transportation buildings and facilities	CIBSE Guide B: 2005 2.3.23; CIBSE AM10: 2005; CIBSE AM13: 2000

Petroleum and Explosives Administration CIBSE Guide B: 2005 2.3.23.3;
and Health and Safety Publication EH40: *Occupational exposure limits for
limiting concentration of exhaust pollutants*. Reference should also be made
to Approved Document B for fire-related issues.

Section 7: Work on existing buildings

When windows are replaced that have trickle ventilators, the replacement
windows should have trickle ventilators sized according to Table 8.7. It is
also recommended that background ventilation should be provided when
other windows are replaced. When windows are replaced as part of a
material change of use, Sections 5 and 6 will also apply.

*Addition of a habitable room (not including a conservatory) to an
existing building*

For background ventilation, if the additional room is connected to a habitable
room, which would now have no windows opening to outside, or it still
has a window(s) but background ventilation with an equivalent area less
than 5000 mm², the guidance for ventilating a habitable room through
another room in Section 5 should be followed.

If the additional room is connected to an existing habitable room that
still has windows opening to outside and with a total background ventilator
equivalent area of at least 5000 mm², there should be background ventilators of at least 8000 mm² equivalent area between the two rooms and
background ventilators of at least 8000 mm² equivalent area between the
additional room and outside.

Alternatively, a single room heat recovery ventilator could be used to
ventilate the additional habitable room. The supply rate to that room

Table 8.7 Background ventilation required when replacing windows

Location	Equivalent area (mm²)
Dwellings	
Habitable rooms	5000
Wet rooms	2500
Buildings other than dwellings	
Occupiable rooms < 10 m²	2500
Occupiable rooms > 10 m²	250 per m² floor area
Kitchens (domestic type)	2500
Bathrooms and shower rooms	2500 per bath or shower
Sanitary accommodation and washrooms	2500 per WC

Note
In all cases, the ventilation opening should not be smaller than was originally provided, and it should
be controllable.

should be determined by calculating (whole building ventilation rate × room volume)/(total volume of all habitable rooms).

For purge ventilation, the guidance in Section 5 on size, location and controls should be followed.

Addition of a wet room to an existing building

Whole building and extract ventilation can be provided by one of the following:

- intermittent extract, as given in Table 8.1, and a background ventilator of at least 2500 mm^2 equivalent area;
- single room heat recovery ventilator;
- passive stack ventilator;
- continuous extract fan.

Addition of a conservatory to an existing building

When adding a conservatory (which is not exempt from the Building Regulation) to an existing habitable room, the guidance in Section 5 on ventilation of a habitable room through a conservatory should be followed with respect to background ventilation, irrespective of the ventilation provisions in the existing room. The purge ventilation should follow the guidance in Section 5 on size, location and controls.

Refurbishing a kitchen or bathroom in an existing dwelling

If 'building work' as defined in the Building Regulations is carried out in a bathroom or kitchen of an existing dwelling the ventilation should not be made worse than before. Existing fans must be retained or replaced. However if there was no extractor fan, PSV or extracting cooker hood in the room before the works, there is no requirement to provide one. The replacement of an existing extract fan or cooker hood with a similar type does not need to be notified to Building Control.

Section 8: Standards and publications

Section 8 consists of a list of standards referred to in the main body of the document and other publications referred to.

Appendix A: Performance-based ventilation

This appendix defines the acceptable level of moisture and pollutants that the ventilation systems in the Approved Document are designed to control.

They may not be adequate to address flueless combustion space heaters, occasional high polluting events, airborne spread of infection, or contamination from other sources. The appendix lists the performance criteria for dwellings and buildings other than dwellings, giving a brief background of how and why they were selected, and finishes with a list of references.

Appendix B: Purge ventilation

The appendix gives a simplification of guidance in BS 5925: 1991 *Code of practice for ventilation principles and designing for natural ventilation*. It gives guidance on opening areas and degrees of opening for windows and doors to provide sufficient purge ventilation. In simple terms, where a window opens at least 30°; the opening area should be a least 1/20 of the room floor area, or 1/10 of the floor area where the opening angle is 15°–30°. External doors can be included in purge ventilation opening area calculations.

Appendix C: Example calculations for ventilation sizing for dwellings

The appendix contains eight examples of ventilation sizing calculations for different types of system.

Appendix D: Minimizing ingress of external pollution into buildings in urban areas

Appendix D gives examples of typical urban pollutants and emission sources and gives guidance on the location and control of ventilation intakes. Location of exhaust outlets is also critical to prevent pollutants re-entering a building or entering adjacent buildings, and guidance is given on their location.

Chapter 9

Approved Document G: Sanitation, hot water safety and water efficiency

The 2010 Edition of Approved Document G has undergone significant changes to take account of additional requirements going beyond the original heading of hygiene. Requirement G1 first deals with cold water supply including the need for **wholesome water** for drinking, food preparation and washing. To ensure compliance with Regulation 36, Water Efficiency in New Dwellings, Requirement G2 has been added. Requirement G3 relates to **heated wholesome water,** including hot water storage, G4 and G5 to sanitary accommodation, washing facilities and bathrooms with G6 added to cover kitchen and food preparation areas.

A number of key terms are defined within the general guidance section of the Approved Document, the majority of which are included in the following text or within other chapters. **Wholesome water** means water complying with the requirements of regulations made under Section 67 (Standards of wholesomeness) of the Water Industry Act 1991, see also Annex 1 of the Approved Document. Chapter 1 outlines where Part G applies to a **material change of use,** the application of the competent person self-certification scheme and non-notifiable work. Where shared hot and/or cold water supplies are provided to Class VI and VII buildings, i.e. domestic greenhouses and conservatories, Requirements G1 and G3 will apply.

REQUIREMENT G1: COLD WATER SUPPLY

(1) There must be a suitable installation for the provision of:
  (a) wholesome water to any place where drinking water is drawn off;
  (b) wholesome water or softened wholesome water to any washbasin or bidet provided in or adjacent to a room containing a sanitary convenience;
  (c) wholesome water or softened wholesome water to any washbasin, bidet, fixed bath and shower in a bathroom; and
  (d) wholesome water to any sink provided in any area where food is prepared.

(2) There must be a suitable installation for the provision of water of suitable quality to any sanitary convenience fitted with a flushing device.

Section G1: Cold water supply

The simple aim of this requirement is to provide buildings with a reliable and wholesome water supply, under sufficient pressure and flow rate for the operation of sanitary appliances, avoiding waste, misuse, excessive consumption and contamination of the water. Wholesome water is not essential for all sanitary appliances; toilets and washing machines for example. Alternative sources of water can be used, subject to a risk assessment for dwellings, and include:

- water abstracted from wells, springs, bore-holes or water courses;
- harvested rainwater or reclaimed **greywater**, see Market Transformation Programme (MTP) 2007; and
- reclaimed industrial process water.

REQUIREMENT G2: WATER EFFICIENCY

Reasonable provision must be made by the installation of fittings and fixed appliances that use water efficiently for the prevention of undue consumption of water.

Requirement G2 applies only when a dwelling is:

(a) erected; or
(b) formed by a material change of use of a building within the meaning of Regulation 5(a) or (b).

Section G2: Water efficiency

For each new dwelling, including flats within a block, the estimated water consumption should not be more than 125 litres/head/day (l/h/d). This should take account of alternative sources of water used and be calculated in accordance with:

- the Water Efficiency Calculator for New Dwellings (downloaded from www.planningportal.gov.uk/uploads/br/water_efficiency_calculator.pdf); or meet
- the minimum water efficiency standard in the 'Code for Sustainable Homes'.

A notice to verify this must be issued to the Building Control Body within five days of completion. Table 1 from the Water Efficiency Calculator is reproduced below (Table 9.1); supporting tables are used for multiple fittings, water softener consumption, greywater and rainwater demand/collection.

Table 9.1 The water calculator for new dwellings

Installation type	Unit of measure	Capacity/ flow rate (1)	Use factor (2)	Fixed use (litres/ person/ day) (3)	Litres/person/ day = [(1) × (2)] + (3) (4)
WC (single flush)	Flush volume (litres)		4.42	0.00	
WC (dual flush)	Full flush volume (litres)		1.46	0.00	
	Part flush volume (litres)		2.96	0.00	
WCs (multiple fittings)	Average effective flushing volume (litres)		4.42	0.00	
Taps (excluding kitchen/ utility room taps)	Flow rate (litres/ minutes)		1.58	1.58	
Bath (where shower also present)	Capacity to overflow (litres)		0.11	0.00	
Shower (where bath also present)	Flow rate (litres/ minutes)		4.37	0.00	
Bath only	Capacity to overflow (litres)		0.50	0.00	
Shower only	Flow rate (litres/ minutes)		5.60	0.00	
Kitchen/ utility room sink taps	Flow rate (litres/ minutes)		0.44	10.36	
Washing machine	Litres/kg dry load		2.1	0.00	
Dishwasher	Litres/ place setting		3.6	0.00	
Waste disposal unit	Litres/use	If present = 1 If absent = 0	3.08	0.00	
Water softener	Litres/ person/day (5)		1.00	0.00	
		Total calculated use (litres/person/day) = (sum column 4)			

continued . . .

Table 9.1 continued ...

Installation type	Unit of measure	Capacity/ flow rate (1)	Use factor (2)	Fixed use (litres/ person/ day) (3)	Litres/person/ day = [(1) × (2)] + 3 (4)
	(6)	Contribution from greywater (litres/ person/day) from Table 4.6			
	(7)	Contribution from rainwater (litres/ person/day) from Table 5.5			
	(8)	Normalisation factor			0.91
	(9)	Total water consumption ('Code for Sustainable Homes') = [(5) – (6) – (7)] × (8) (litres/person/day)			
	(10)	External water use			5.0
	(11)	Total water consumption (Building Regulation 37) = (9) + (10) (litres/ person/day)			

REQUIREMENT G3: HOT WATER SUPPLY AND SYSTEMS

1. There must be a suitable installation for the provision of heated whole-some water or heated softened wholesome water to:

   (a) any washbasin or bidet provided in or adjacent to a room containing a sanitary convenience;

   (b) any washbasin, bidet, fixed bath and shower in a bathroom; and

   (c) any sink provided in any area where food is prepared.

2. A hot water system, including any cistern or other vessel that supplies water to or receives expansion water from a hot water system, must be designed, constructed and installed so as to resist the effects of temperature and pressure that may occur either in normal use or in the event of such malfunctions as may reasonably be anticipated, and must be adequately supported.

3. A hot water system that has a hot water storage vessel must incorporate precautions to:

   (a) prevent the temperature of the water stored in the vessel at any time execeeding 100°C;

   (b) ensure that any discharge from safety devices is safely conveyed to where it is visible but will not cause a danger to persons in or about the building.

   Requirement G3(3) does not apply to a system which heats or stores water for the purposes only of an industrial process.

4. The hot water supply to any fixed bath must be so designed and installed as to incorporate measures to ensure that the temperature of the water that can be delivered to that bath does not exceed 48°C.

   Requirement G3(4) applies only when a dwelling is:

   (a) erected; or

   (b) formed by a material change of use of a building within the meaning of Regulation 5(a) or (b).

Section G3: Hot water supply and systems

The extended scope of this requirement seeks to provide buildings with **heated wholesome water** to sanitary appliances, avoiding waste, misuse and excessive consumption. The application is limited to:

- washbasin provided for toilet accommodation, Requirement G4(2);
- washbasin, bidet, fixed bath or shower in residential accommodation, Requirement G5; and
- sink in a food preparation area, Requirement G6.

The utilization of hot water storage systems gives rise to potential health risks to persons in or about buildings. The storage vessel itself is subject to high internal pressures, and water discharged via safety devices will be very hot. The guidance, which also includes other forms of water heating, is split into a number of headings that can be summarized below.

General

- Hot water storage systems should accord with BS 6700: 2006 + Amendment 1: 2009 *Specification for design, installation, testing and maintenance of services supplying water for domestic use within buildings and their curtilages* or BS EN 12897: 2006 *Water supply. Specification for indirectly heated unvented (closed) storage water heaters.* The vessels themselves to accord with BS 853–1: 1996, BS 1566–1: 2002 or BS 3198: 1981 as appropriate.

Vented hot water storage systems

*   to incorporate a vent pipe, minimum 19 mm diameter, discharging to a cold water storage cistern;
*   for direct heat sources – incorporate a non-self-resetting energy cut-out;
*   for indirect heat sources – incorporate an overheat cut-out and a temperature relief valve, or combined temperature and pressure relief valve.

Unvented hot water storage systems

*   In addition to any thermostat, to incorporate a non-self-resetting energy cut-out and a temperature relief valve, or combined temperature and pressure relief valve.
*   Systems up to 500 litres capacity and 45 kW power output (or power input more than 45 kW) should take the form of a **propriety unit or package** and be indelibly marked with information to include: manufacturer; model; rating; operating pressure/data; and a warning to the user/installer.
*   Systems over 500 litres capacity and over 45 kW power input will generally be designed on an individual basis by a suitably qualified engineer with a competent person used to install the system.

Discharge pipes

*   The typical discharge pipe arrangements from a safety device should include a tundish within 600 mm of the device, for visual warning, and its termination to a safe location, e.g. a trapped gulley, avoiding asphalt or felt roofs and non-metallic rainwater goods.
*   For specific design guidance reference should be made to Diagram 1 and Table 1 of the Approved Document.

Electric water heating

*   Electric fixed immersion heaters, instantaneous water heaters and storage water heaters should accord with the relevant part of BS EN 60335.

Specification for safety of household and similar electrical appliances. Solar water heating

- Factory made systems should accord with BS EN 12976–1: 2006 *Thermal solar systems and components*. Other systems should use prEN/TS 12977–1: 2008, BS 5918: 1989 or *CIBSE Guide G*.
- Where these systems are used an additional heat source should be available, this may also be necessary to maintain temperatures to avoid microbial growth.

Prevention of scalding

For all new dwellings, including those formed under a change of use, precautions must be taken to limit the temperature of hot water supplied to fixed baths to a maximum of 48 °C. To achieve this, in-line blending valves, which cannot be easily altered, should be used. For guidance reference should be made to BRE Information Paper 14/03 *Preventing hot water scalding in bathrooms: using TMV's* and to *NHS Estates Model specification D 08* for care homes and similar establishments. In addition the temperature of hot water supplied to the domestic distribution system should not exceed 60 °C.

REQUIREMENT G4: SANITARY CONVENIENCES AND WASHING FACILITIES

1. Adequate and suitable sanitary conveniences must be provided in rooms provided to accommodate them or in bathrooms.
2. Adequate hand washing facilities must be provided in:

   (a) rooms containing sanitary conveniences; or
   (b) rooms or spaces adjacent to rooms containing sanitary conveniences.

3. Any room containing a sanitary convenience, a bidet, or any facility for washing hands provided in accordance with sub-paragraph (2)(b), must be separated from any kitchen or any area where food is prepared.

Section G4: Sanitary conveniences and washing facilities

The wording of Requirement G4 is self-explanatory, seeking to require satisfactory and sufficient toilet accommodation with associated washing provisions to *all* building purpose groups. A summary of the Approved Document guidance is as follows.

- A dwelling, including a house or flat, should have a minimum of one WC and one washbasin.

- Only a door, not a lobby, is needed to separate a WC or urinal from a food preparation area, including a kitchen.
- The washbasin should be located in the same room as the WC or in an adjoining room or area, subject to this not being used for food preparation. The washbasin should have a hot water supply (either direct from a central source or an instantaneous water heater) and discharge via a trap to the foul water drainage system.
- A flushing WC or urinal should discharge via a trap to the foul water drainage system.
- A WC utilizing a macerator, pump and small-bore pipe can discharge to the foul water drainage subject to the availability of a traditional WC connected by gravity to the foul water drainage system.
- Chemical or composting WCs and urinals may be used where suitable arrangements exist for disposal of the waste, on or off the site, without passing through any living space or food preparation area.

For the number, type and sitting of appliances, in buildings other than dwellings and in addition to the above, reference should be made to:

- the Workplace (Health, Safety and Welfare) Regulations 1992 (supported by the Approved Code of Practice);
- Approved Document M – Access and facilities for disabled people.

- BS 6465: Part 1: 2006 and Amendment 2009 *Code of Practice for the design of sanitary facilities and scales of provision of sanitary and associated appliances.* This British Standard incorporates minimum requirement tables for a range of building purpose groups, including dwellings, residential homes, offices and shops, factories, schools, assembly buildings and hotels.

REQUIREMENT G5: BATHROOMS

A bathroom must be provided containing a washbasin and either a fixed bath or a shower.

Requirement G5 applies only to dwellings and to buildings containing one or more rooms for residential purposes.

Section G5: Bathrooms

The Approved Document guidance is straightforward, reflecting the self-explanatory wording of the requirement. It is important to note that bathroom provisions apply *only* to dwellings and rooms for residential purposes, and reference will therefore need to be made to the legislation listed under Requirement G4 concerning bath and shower facilities in offices and shops, factories, hotels, etc.

- a dwelling, including a house or flat, should have a minimum of one bathroom with a fixed bath or shower and washbasin;
- the bath or shower should have a hot water supply, hot tap on the left, and discharge via a trap to the foul water drainage system;
- a bath or shower utilizing a macerator, pump and small-bore pipe can discharge to the foul water drainage subject to the availability of traditional washing facilities connected by gravity to the foul water drainage system.

REQUIREMENT G6: KITCHENS AND FOOD PREPARATION AREAS

A suitable sink must be provided in any area where food is prepared.

Section G6: Food preparation areas

Quite simply a sink should be provided in any kitchen or food preparation area in all buildings. An additional sink is not required in a separate room where a dishwasher is located. Finally, separate hand washing facilities may also be required where the Food Hygiene (England) Regulations 2006 apply.

Chapter 10

Approved Document H: Drainage and waste disposal

The satisfactory drainage of a building and the efficient disposal of refuse are both matters that could have a dramatic influence on public health and safety. A blocked foul or surface water drain, for example, could cause effluent leakage or flooding; a badly designed septic tank could allow leakage into the subsoil or watercourse; and insufficient dustbin provision could give rise to rodent infestation. It is therefore important that compliance with the functional Requirements H1–H4, is shown by referring to the guidance contained in Approved Document H.

REQUIREMENT H1: FOUL WATER DRAINAGE

1. An adequate system of drainage shall be provided to carry foul water from appliances within the building to one of the following, listed in order of priority:

   (a) a public sewer, or, where that is not reasonably practicable;
   (b) a private sewer communicating with a public sewer, or, where that is not reasonably practicable;
   (c) either a septic tank, which has an appropriate form of secondary treatment or another wastewater treatment system, or, where that is not reasonably practicable;
   (d) a cesspool.

2. In this part 'foul water' means wastewater, which comprises or includes:

   (a) waste from a sanitary convenience, bidet or appliance used for washing receptacles for foul waste; or
   (b) water that has been used for food preparation, cooking or washing.

Requirement H1 does not apply to the diversion of water that has been used for personal washing or for the washing of clothes, linen or other articles to collection systems for reuse.

To comply with the Requirement the foul water drainage system should:

- convey the foul water flow to a suitable outfall;
- minimize the risk of leakage and blockages;
- prevent foul air from entering the building;
- be ventilated and made accessible for blockage clearance; and
- not increase the vulnerability of the building to flooding.

An important point to note is that the Requirement seeks to control the foul water drainage system *only* and *not* the suitability of the outfall itself, where guidance from the local water authority should be sought.

Before outlining the guidance contained in Sections 1 and 2 of the Approved Document flow rate data may be necessary to verify that the capacity of the system will be sufficient for the expected flow at any point in the system. Reference can therefore be made to Table 10.1.

Section 1: Sanitary pipework

Figure 10.1 illustrates the principal provisions for sanitary pipework for domestic buildings and small non-domestic buildings. Refer to guidance in Appendix A of Approved Document for larger buildings and BS EN 12056 for complex designs (including where WC major flush volumes are less than 5 litres). The specific provisions are then discussed under each element of the above ground drainage system starting with the need to provide a water seal (or trap) at all points of discharge into the system.

Traps

- should prevent foul air in the system entering the building by retaining, under working/test conditions, a minimum seal of 25 mm;
- should be removable, have a cleaning eye or come away with the appliance.

Table 10.1 Flow rates

Number of dwellings (1 WC, 1 bath, 1 sink and 1 or 2 washbasins)	Flow rate (l/s)	Individual appliances (l/s)	Flow rate (l/s)
1	2.5	Washdown WC	2.3
5	3.5	Urinal (per person unit)	0.15
10	4.1	Washbasin (32 mm branch)	0.6
15	4.6	Sink (40 mm branch)	0.9
20	5.1	Bath (40 mm branch)	1.1
25	5.4	Auto washing machine	0.7
30	5.8	Shower	0.1
		Spray tap basin	0.06

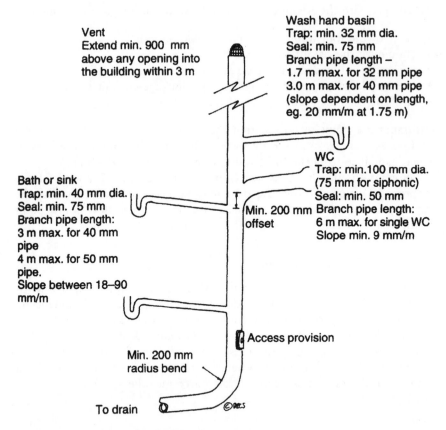

Figure 10.1 Discharge stack (S&VP) and branch pipes.

Note
For unvented (common) branch discharge pipes serving more than one appliance apply the following:
WCs – up to eight in number or 15 m branch length, 100 mm pipe size
Bowl urinals – up to five (branch as short as possible), 50 mm pipe size
Stall urinals – up to seven (branch as short as possible), 65 mm pipe/trap size
Washbasins – up to four or 4 m branch length (no bends), 50 mm pipe size.

Branch discharge pipes

- Should discharge to another branch discharge pipe or a discharge stack, unless the appliances are on the ground floor, where they may discharge to a stub stack, gulley (waste water only) or directly to the drain.
- Should not discharge into a stack lower than 450 mm above the drain invert level, for single dwellings up to three storeys.
- Should not discharge into a stack lower than 750 mm above the drain invert level, for buildings up to five storeys; for buildings over five

storeys [and those over 20 storeys] connect ground [ground and first] floor appliances to a separate stack (gulley or drain if applicable).

- Ground-floor WCs may discharge directly to the drain where the maximum dimension above drain invert level is 1.3 m.
- Pipe entry into a gulley should be above the water seal but below the grating or access plate.
- Avoid bends or utilize the largest radius possible.
- If the length and slope exceeds the figures given in Figure 10.1 then to avoid loss of trap seals due to system pressures the branch pipe should be ventilated to the external air by a branch ventilation pipe, or via the discharge stack (modified single stack system) or via a ventilating stack (ventilated system).
- Branch ventilating pipes should: connect within 300 mm of the trap and above spillover level (of the highest appliance) at stack; terminate as discharge stack (Figure 10.1) and be a minimum 25 mm diameter (35 mm diameter if longer than 15 m or pipe has more than five bends).
- Unless removable traps (or appliances) can be used then lengths of pipe should have rodding access.
- Condensate drainage from boilers may connect to sanitary pipework using minimum 22 mm diameter pipe via 75 mm condensate trap.

Discharge stacks

- Minimum diameters: 50 mm (no WCs) up to 1.2 l/s; 65 mm (no WCs) up to 2.1 l/s; 75 mm (one siphonic WC) up to 3.4 l/s; 90 mm up to 5.3 l/s; 100 mm up to 7.2 l/s.
- Offsets in the 'wet' portion should be avoided, but they may be used in buildings up to three storeys if no branch connection is made within 750 mm. In higher buildings a separate ventilating stack may be needed with connections above and below offset.
- Stacks should be inside the building if more than three storeys high.
- To avoid loss of trap seals due to system pressures the stack should be ventilated to the external air, terminated as in Figure 10.1. This may be reduced to 75 mm diameter for houses up to two storeys, minimum 50 mm in other cases if the drain is liable to surcharging. Consideration could also be given to the use of an **air admittance valve** located within the building and subject to a current Agrément Certificate.
- Allow reasonable access provisions for rodding purposes and repair of system pipework.

Stub stacks

- An unventilated stub stack may connect above ground to a ventilated discharge stack or to a drain not subject to surcharging.

- No branch should connect more than 2 m above the invert level of the drain (or discharge stack connection), 1.3 m maximum for a WC.
- Length of branch drain serving a stub stack should be a maximum of 6 m for a single appliance and 12 m for a group of appliances (unless ventilated).

A range of materials may be used for sanitary pipework, including cast iron, copper, galvanized steel, uPVC, polypropylene and plastics (some may not however be suitable for trade effluent or condensate from boilers). Positive fixing methods should be adopted that allow for thermal movement of the pipes. Consideration may also need to be given to the separation of different metals to avoid electrolytic corrosion. Once installed, the pipes, fittings and joints should withstand an air test (or smoke test, but not suitable for uPVC pipes) of positive pressure, minimum 38 mm water gauge, for at least 3 minutes; traps to retain a minimum 25 mm seal.

An **alternative approach** to the guidance above is to refer to BS EN 12056 *Gravity drainage systems inside buildings*. See also BS EN 12109 *Vacuum drainage systems inside buildings*.

Section 2: Foul drainage

The below-ground foul water drainage system comprises the necessary pipes and fittings to connect sanitary pipework (i.e. discharge stacks, stub stacks and gullies) to the outfall (i.e. a public or private sewer, cesspool or septic tank). Foul water drains may discharge to a combined public sewer, which carries both foul and surface water, although the pipe sizes may need to be increased to take account of both flow rates. The system layout should be kept as simple as possible, should minimize changes of direction and gradient, and should provide for access points where blockages could not be cleared without them. The specific provisions are discussed under headings reflecting the components of the system.

Outlets

- Connect to a public foul or combined sewer wherever this is reasonably practicable. For small developments connection, gravity or pumped, should be made to a public sewer where this is within 30 m, subject to ownership of the land. A developer may requisition a sewer from the sewerage undertaker, who has powers to construct sewers over private land, where serving more than one curtilage. Notify sewerage undertaker at least three weeks before any sewer connection is made.
- If the above is not possible, connect to an existing private sewer, in good condition with sufficient capacity, which connects with a public sewer, subject to appropriate permissions.

- The final alternative is to use a wastewater treatment system or cesspool.

Surcharging of drains

- Consult the sewerage undertaker if low-lying sites are at increased risk of flooding.
- For a high-risk basement, containing sanitary appliances, the drainage should be pumped. For a low-risk situation an anti-flooding valve should be installed.
- For other low-lying/low-risk sites (not basements) provide an overflow gully, at least 75 mm below the floor level. An anti-flooding valve (with suitable notices) should be provided, or the drainage system pumped, for high-risk areas.
- All drainage unaffected by surcharge should bypass the protective measures and discharge by gravity.

Layout

- Connection of drains to other drains or private or public sewers and of private sewers to public sewers should be made obliquely, or in the direction of flow, and with prefabricated components if possible. Use suitable repair couplings where connecting to existing drain.
- Locate sewers (serving more than one property) to avoid future extensions.
- Drainage serving kitchens in commercial hot food premises should be fitted with a grease separator complying with prEN 1825–1, designed to prEN 1825–2, or other effective means of grease removal.
- Sewers (drain serving more than one property) should normally have a minimum diameter of 100 mm for up to 10 dwellings, minimum diameter of 150 mm where serving more than 10 dwellings.

Pipes

- The drainage pipes should have sufficient capacity to carry the anticipated flow and be laid to falls. Table 10.2 gives specimen values.
- Any change of gradient should incorporate an access point.
- Pipes should be laid in straight lines, or slight curves if blockages can still be cleared. Bends used should be close to an inspection chamber or manhole and to the foot of discharge or ventilating stacks.
- The system should be ventilated at or near the head of each main drain, any branch drain longer than 6 m serving a single appliance or 12 m serving a group of appliances, and to a drain fitted with an intercepting trap. A ventilating discharge stack would achieve this.

Table 10.2 Minimum gradients and maximum capacities for foul water drains

Pipe diameter (mm)	Gradient	Flow rate (l/s)
100	1:20	13.0*
100	1:40	9.2 (at peak flow rate <1 l/s)
100	1:60	7.4*
100	1:80 (min. 1 WC)	6.3 (at peak flow rate >1 l/s)
150	1:40	26.0*
150	1:60	21.0*
150	1:80	19.0*
150	1:100	17.0*
150	1:150 (min. 5 WCs)	15.0 (at peak flow rate >1 l/s)

Note

* Denotes approximate values taken from Approved Document guidance, for foul drains running 0.75 proportional depth. Minimum pipe diameters: 75 mm for waste water and 100 mm for soil water or trade effluent.

- Where pipes run under or near a building, on piles/ground beams, in common trenches or in unstable ground, precautions may be necessary to avoid settlement of drain runs. See also TRL *A guide to the design loadings for buried rigid pipes*.
- For drain runs under buildings adopt a minimum 100 mm granular or other flexible filling around the pipe. Where excessive settlement may occur a suspended drainage system or other solution may be necessary.
- A pipe within 300 mm of the underside of a slab should be encased in concrete and made integral with the slab.
- Where pipes need to pass through a wall or foundation then a sleeve (minimum 50 mm clearance) or rocker pipes should be used to retain drain flexibility at these points.
- A drain trench within 1 m of a building should be backfilled with concrete up to the foundation soffit level. Where 1 m or more from the building the trench should be filled with concrete to within that dimension to the foundation soffit (less 150 mm).
- Consideration should also be given to where rodent infestation may cause a problem. The Local Authority should be able to confirm the extent of any problem and precautions to be taken; these include: sealed drainage; intercepting traps; rodent barriers; cages to vent stacks and covers/gratings to gullies.
- Where a sewer connection by gravity is impracticable a pumped system may be used; see BS EN 12050, 12056–4 or 752–6. Allow for 24 hour chamber capacity at 150 litres/head/day.

Bedding and backfilling

- Choice will be dependent on the pipe depth below ground level, the size and strength of the pipe, and the extent (or weight) of backfilling over the pipe.
- For **rigid pipes** of standard strength adopt 100 mm granular fill bedding and 150 mm cover of selected backfill. For detailed provisions refer to Approved Document guidance. If pipes have less than the cover recommended in Table 8 then minimum 100 mm concrete encasement may be necessary, with movement joints at each pipe socket.
- For **flexible pipes** adopt 100 mm granular fill bedding and surround with 300 mm cover of selected backfill, as described in the Approved Document. The minimum depth of a drain under a road should be 1200 mm (unless bridged or encased in reinforced concrete) and 600 mm under fields or gardens (unless bridged with paving slabs), maximum depth 10 m.

Blockage clearance

- The provision of access points to clear blockages assumes that traditional methods of rodding are used (which need not be in the direction of flow) and not mechanical or other means of clearing, which may be justifiable in certain circumstances.
- Access points are either: a rodding eye (capped extension of the drain); an access fitting (with no open channel), an inspection chamber (ground-level working space); or a manhole (with drain-level working space). The minimum dimensions are listed in Table 10.3, noting that due allowance should be made to accommodate all branch connections.
- Access points should be sited: at or near the head of each drain run; at a bend or change of gradient; at a change of pipe size; and at a junction (unless clearance is possible from an access point, noting that rodding may only be possible from one direction).

A range of materials may be used for pipes and access points: clay, concrete or grey iron for rigid pipes; uPVC for flexibly jointed pipes; and brick, concrete or plastic for access points. Where possible, flexible joints should be adopted, which allow for any differential settlement of the pipes. Consideration may also need to be given to the separation of different metals to avoid electrolytic corrosion. Once laid, the drains and access points should exclude groundwater and rainwater and withstand a water test filled to 500 mm above the drain invert. Alternatively an air test can be used to ensure a maximum loss of head on a manometer of 25 mm in 7 minutes for a 100 mm gauge (12 mm for a 50 mm gauge). Also refer to Regulation 18 concerning the testing of drains. An **alternative approach** to the guidance

Table 10.3 Minimum dimensions for access points

Type	Depth to (m)	Internal sizes (mm)	Spacing (m) to IC	MH
Rodding eye	–	Minimum 100	45	45
Small access fitting	0.6 or less	150 × 100 (150 dia.)	22	22
Large access fitting	0.6 or less	225 × 100	22	45
Inspection chamber (IC)	0.6 or less	190 dia. (max 150 drain)	45	45
	1.2 or less	450 × 450 (450 dia.)	45	45
Manhole (MH)	1.5 or less	1200 × 750 (1200*† dia.)	45	90
[600 × 600 mm or over 600 mm dia. cover size]	over 1.5	1200 × 1075 (1200 dia.)*	45	90
Shaft	over 3.0	1050 × 800 (1050 dia.)	45	90

Notes
Covers to ICs and MHs should be removable and non-ventilating, of durable material (e.g. cast iron, steel or uPVC), and of suitable strength, bearing in mind location. MHs deeper than 1 m should have step irons or ladder to allow for access.
* For 300 mm largest pipe.
† Minimum size of manhole serving a sewer.

above is to refer to BS EN 752 and BS EN 12056; this includes a discharge unit method of determining pipe sizes.

Additional issues now covered in the latest Approved Document include:

- **Greywater recovery systems** – Sanitary pipework/underground drainage used to collect greywater for recovery and reuse should follow the Approved Document guidance, and be clearly marked. For further information refer to Water Regulations Advisory Scheme leaflet No. 09–02–04 *Reclaimed water systems. Information about installing, modifying or maintaining reclaimed water systems.*
- Guidance is offered, in Appendix H1-B of the Approved Document, on the other legislation applicable to repairs, alterations and discontinued use of drains and sewers.
- Appendix H1-C of the Approved Document deals with the adoption of sewers and connection to public sewers.

As an alternative to a public or private sewer connection, which may not be available, a cesspool, septic tank or small sewage treatment works may be used. Cesspools are basically large enclosed effluent storage tanks, which are emptied periodically. Whereas septic tanks treat the sewage internally and discharge it to the subsoil via a system of land drains or a filter bed, desludging is normally required about once a year. Septic tanks are generally economic for up to three dwellings. The design and installation of a small

REQUIREMENT H2: WASTEWATER TREATMENT SYSTEMS AND
CESSPOOLS

1. Any septic tank and its form of secondary treatment, other wastewater
   treatment system or cesspool, shall be so sited and constructed that:

   (a) it is not prejudicial to the health of any person;
   (b) it will not contaminate any watercourse, underground water or water
       supply;
   (c) there are adequate means of access for emptying and maintenance;
       and
   (d) where relevant, it will function to a sufficient standard for the
       protection of health in the event of a power failure.

2. Any septic tank, holding tank that is part of a wastewater treatment
   system or cesspool shall be:

   (a) of adequate capacity;
   (b) so constructed that it is impermeable to liquids; and
   (c) adequately ventilated.

3. Where a foul water drainage system from a building discharges to a
   septic tank, wastewater treatment system or cesspool, a durable notice
   shall be affixed in a suitable place in the building containing information
   on any continuing maintenance required to avoid risks to health.

sewage treatment system should comply with the recommendations of
BS 6297: 1983. The main provisions and alternatives are illustrated in
Figure 10.2.

The design considerations for cesspools and septic tanks can be
summarized as follows:

- They should have sufficient capacity, below the inlet level, as indicated
  in Figure 10.2, sited at least 7 m from any habitable part of buildings.
- They should be located to prevent the emptying, desludging and cleaning
  operations from causing a hazard to the building occupants or the
  removal of contents through a dwelling or work space. Access may be
  via an open covered space.
- They should prevent leakage of contents and the ingress of subsoil
  water.
- They should be constructed of brickwork (minimum 220 mm
  engineering bricks in 1:3 mortar), in situ concrete (minimum 150 mm
  thick of C/25/P mix), or glass-reinforced plastic, polyethylene or steel,
  complying with BS EN 12566-1.

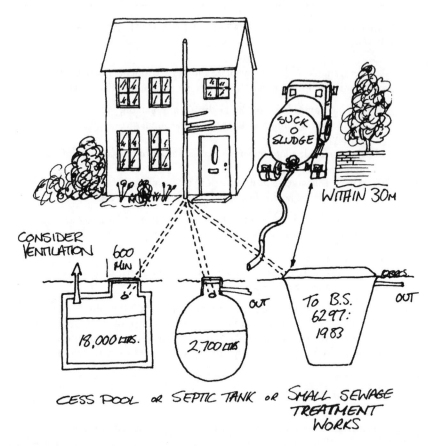

Figure 10.2 Alternatives for sewage disposal.

- They should have suitable concrete covers and ventilation (septic tanks and settlement tanks may be open and fenced in).
- Where covered, access provision should be made for emptying, desludging and cleaning. Access covers should be of durable material (noting contents), minimum 600 mm dimension and be lockable.
- The inlet to a cesspool and both the inlet and outlet to a septic tank or settlement tank should have access provision for inspection.
- To minimize turbulence in a septic tank the inlet flow rate should be limited: for example, by the use of a dip pipe (for a tank up to 1200 mm in width) or a 1:50 fall for the last 12 m of the drain run (drains up to 150 mm).

- Provide a notice within the building describing the system and necessary maintenance.
- Any discharge from a wastewater treatment system is likely to require a consent from the Environment Agency. This would normally be undertaken at the Town Planning stage, and certainly before the Building Regulation approval submission. Reference can be made to the Environment Agency's Pollution Prevention Guideline No. 4 *Disposal of sewage where no mains drainage is available.*
- Drainage fields or drainage mounds, acting as land drains/filter beds, are used to provide secondary treatment to the discharge from a septic tank or small sewage treatment works/package plant.
- Drainage field or mound should be located at a minimum: 10 m from any watercourse or permeable drain; 15 m from any building; 50 m from the point of abstraction of any groundwater supply; not in any zone 1 groundwater protection zone; and isolated from any other drainage field, mound or soakaway.
- The Approved Document offers detailed guidance on ground conditions, percolation testing and design/construction.
- An alternative to drainage fields/mounds, now fully detailed in the Approved Document, is constructed wetlands/reed beds. There are two main designs of constructed wetland system, horizontal flow and vertical flow. Guidance on maintenance requirements for reed bed treatment systems is given in BRE Good Building Guide No. 42.
- Greywater or rainwater tanks (not water butts) should prevent leakage of the contents and ingress of subsoil water, be ventilated, have an anti-backflow device where applicable and have access provisions for emptying and cleaning. Refer to the Water Regulations Advisory Scheme leaflet No. 09–02–04, *Reclaimed water systems. Information about installing, modifying or maintaining reclaimed water systems,* for more information.
- For small sewage or packaged treatment works the discharge from the wastewater treatment plant should be sited at least 10 m away from watercourses and any other buildings. Packaged treatment works should be type-tested to BS 7781 or by notified body.
- Appendix H2-A deals with maintenance of wastewater treatment systems or cesspools and explains the background legislation and powers of the Environment Agency.

The **alternative approach** to comply with Requirement H2 is to follow the relevant recommendations of BS 6297: 1983 *Code of practice for design and installation of small sewage treatment works and cesspools.* Formulae are included for the sizing of tanks.

REQUIREMENT H3: RAINWATER DRAINAGE

1. Adequate provision shall be made for rainwater to be carried from the roof of the building.
2. Paved areas around the building shall be so constructed as to be adequately drained.
3. Rainwater from a system provided pursuant to sub-paragraphs (1) or (2) shall discharge to one of the following, listed in order of priority:

   (a) an adequate soakaway or some other adequate infiltration system, or, where that is not reasonably practicable;
   (b) a watercourse, or, where that is not reasonably practicable;
   (c) a sewer.

Requirement H3(2) applies only to paved areas:

(a) that provide access to the building pursuant to paragraph M2 of Schedule 1 (access for disabled people);
(b) that provide access to or from a place of storage pursuant to paragraph H6(2) of Schedule 1 (solid waste storage); or
(c) in any passage giving access to the building, where this is intended to be used in common by the occupiers of one or more other buildings.

Requirement H3(3) does not apply to the gathering of rainwater for reuse.

The wording of the expanded Requirement is quite clear in that the surface water drainage system should:

- convey the roof water flow to a suitable outfall; avoiding damage to foundations;
- minimize the risk of leakage and blockages;
- be made accessible for blockage clearance.

As with foul water drainage it is important to note that the requirement seeks to control the drainage system up to the outfall *only* and *not* the suitability of the outfall itself. Guidance from the local water authority may be necessary in this regard. For example, an existing surface water sewer may not be of a sufficient capacity to accept the proposed flow rates from the building; means to retain and restrict water outflow could therefore be required. Figure 10.3 serves to illustrate the provisions.

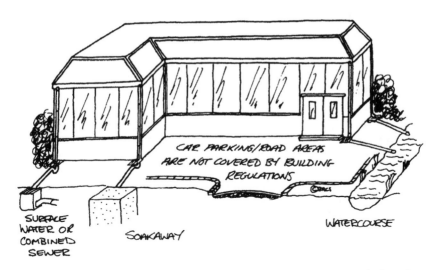

Figure 10.3 Roof drainage disposal provisions. The drainage of small individual roof areas of 6 m² or less is not controlled.

Section 1: Gutters and rainwater pipes

To calculate the effective roof area(s) to be drained the following formula can be used:

Effective design area (m^2) = plan area × multiplication factor (MF)

where the MF for a flat roof = 1, the MF for a pitched roof at 30° = 1.29, at 45° = 1.50, at 60° = 1.87, and at over 70° or any wall = 0.50.

Once the roof area to be drained has been established, the gutter and outlet (or downpipe) sizes can be found from Table 10.4. For more detailed design guidance reference should be made to BS EN 12056.

Table 10.4 Gutter and outlet sizes

Max effective roof area (m^2)	Gutter size (mm dia.)	Outlet size (mm dia.)	Flow capacity (l/s)
6	–	–	–
18	75	50	0.38
37	100	63	0.78
53	115	63	1.11
65	125	75	1.37
103	150	89	2.16

Note
Table relates to a half-round gutter, laid level, sharp-edged outlet at one end and where the distance to the outlet from a stop end does not exceed 50 times the water depth.

The remaining provisions are as follows.

- Gutters should be laid to allow overflow away from the building.
- Rainwater pipes should discharge to a drain or gulley, or to another gutter or other drained surface (taking care to avoid water flow over a pedestrian route).
- Discharge to a combined drain should be via a trap.
- A rainwater pipe serving more than one roof area should have the equivalent cross-sectional area of all the outlets.
- Gutters should remain watertight, and internal pipes should be able to withstand the air test as described for sanitary pipework.
- Gutters and downpipes should be of durable materials, with adequate strength and fixings to allow for thermal movement. Consideration may also need to be given to the separation of different metals to avoid electrolytic corrosion.
- Syphonic roof drainage systems should be designed in accordance with BS EN 12056–3; also refer to Hydraulics Research Ltd Report SR 463 *Performance of syphonic drainage systems for roof gutters.*
- Eaves drop systems allow rainwater to drop freely to the ground and follow good design practice to avoid water ingress.
- Rainwater recovery systems should have storage tanks complying with H2 guidance and be suitably marked. Also refer to the Water Regulations Advisory Scheme leaflet No. 09–02–04, *Reclaimed water systems. Information about installing, modifying or maintaining reclaimed water systems.*

BS EN 12056 *Gravity drainage systems inside buildings* is referred to as the **alternative approach.**

Section 2: Drainage of paved areas

This section has been added to indicate how compliance can be shown with Requirement H3(2), although its application is limited. Guidance is given for paved areas around buildings and small car parks up to 4000 m^2; over this area refer to BS EN 752–4. Design rainfall intensities can be taken as 0.014 l/s/m^2 or refer to Diagram 2 of the Approved Document. The options to deal with these paved areas are listed below:

- Paths, driveways and other narrow areas can be freedraining to a pervious area, e.g. grass.
- Where suitable a porous or permeable surface can be used, overlying a granular layer. Also refer to CIRIA report C522 *Sustainable urban drainage systems – design manual for England and Wales.*
- Where neither of the above can be used, gullies and/or channels will need to be used, connected to a surface water drainage system. Precautions will need to be taken to prevent silt buildup.

The **alternative approach** is given by way of BS EN 752–4: 1998 *Drain and sewer systems outside buildings, Part 4 Hydraulic design and environmental aspects.*

Section 3: Surface water drainage

As with foul water the below-ground surface water drainage system comprises the necessary pipes and fittings to connect the rainwater pipes to the outfall(s). This should be a soakaway or other infiltration system, or if not practicable, to a sewer. Surface water drains may discharge to a combined public sewer that carries both foul and surface water. Where the public combined sewer (or private drain) cannot accommodate the additional flow, or where required by H5 a separate system with its own outfall should be installed. For all provisions relating to pipes, bedding and backfilling, blockage clearance, materials and testing direct reference should be made to the design guidance contained in Section 2 under Requirement H1.

The guidance in Section 3 relates to the design of surface water drainage systems for small catchments with impervious areas up to 2 ha; refer to BS EN 752–4 for larger catchments. The remaining design considerations are listed below:

- The drainage pipes should have sufficient capacity to carry the anticipated flow from the roof and any runoff from paved areas; Table 10.5 gives specimen values.
- Discharge to a watercourse may require a consent from the Environment Agency, where the discharge rate may be limited.
- Materials or processes, which may cause pollution, should have a separate drainage system with appropriate separator or method of treatment.

Table 10.5 Minimum gradients and maximum capacities for surface water drains

Pipe diameter (mm)	Gradient	Flow rate (l/s)*
75	1:10	9.0
75	1:50	3.9
75	1:100 (minimum fall)	2.8
100	1:10	20.0
100	1:50	8.6
100	1:100 (minimum fall)	6.0
150	1:50	26.0
150	1:100	18.0
150	1:150	16.0

Note
* Denotes approximate values taken from Diagram 1 of the Approved Document, Section 2 of H3, for rainwater drains running full. Minimum pipe diameter 75 mm. Capacity can be increased by increasing fall.

- The guidance now makes reference to car parks, petrol-filling stations and other similar areas where oil interceptors should be provided. Refer to Appendix H3-A of the Approved Document (which also gives the background legislation) and *Use and design of oil separators in surface drainage systems*, Pollution Prevention Guideline No. 3.
- Infiltration devices, including soakaways, swales, infiltration basins and filter drains, should only be used where the subsoil conditions allow. See also CIRIA Report 156 – *Infiltration drainage – manual of good practice*.
- For areas less than 100 m² traditional square or circular soakaway pits can be used. Design guidance is offered in the Approved Document where percolation tests would be required. For further reference see BS EN 752–4 or BRE Digest 365 *Soakaway design*.
- The other types of infiltration system are swales (grass-lined channels); infiltration basins (grass-lined basins that allow water to infiltrate the ground); filter drains or French drains (perforated pipes in gravel bed/surround); and detention ponds (can attenuate the flow).

The requirement can also be met by following the **alternative approach** in BS EN752–4 *Drain and sewer systems outside buildings*.

REQUIREMENT H4: BUILDING OVER SEWERS

1. The erection or extension of a building or work involving the under-pinning of a building shall be carried out in a way that is not detrimental to the building or building extension or to the continued maintenance of the drain, sewer or disposal main.
2. In this paragraph 'disposal main' means any pipe, tunnel or conduit used for the conveyance of effluent to or from a sewage disposal work, which is not a public sewer.
3. In this paragraph and paragraph H5 'map of sewers' means any records kept by a sewerage undertaker under Section 199 of the Water Industry Act 1991 (a).

   (a) 1991 c. 56.

Requirement H4 applies only to work carried out:

(a) over a drain, sewer or disposal main, which is shown on any map of sewers; or
(b) on any site or in such a manner as may result in interference with the use of, or obstruction of the access of any person to, any drain, sewer or disposal main, which is shown on any map of sewers.

This new requirement replaces Section 18 of the Building Act 1984, which is now repealed. Specific performance criteria are applied which reflect the wording of the requirement. Provisions apply to the construction, extension or underpinning of a building over or within 3 m of the centreline of an existing drain, sewer or disposal main shown on the sewerage undertaker's sewer records whether that sewer is a public sewer or not. These maps are held by the sewerage undertaker and local authorities, and are available for inspection during office hours. Consultation procedures still have to take place where the developer should consult the owner of the drain or sewer. The owner of any public sewer is the sewerage undertaker; they would undertake any repair/replacement to their own sewers and offer advice as to the condition of their sewers. Where a building may be constructed over a sewer intended for adoption, the sewerage undertaker should be consulted. The Building Control Body also has a duty to consult the sewerage undertaker, at the time of the building regulation submission, and allow them 15 days to comment. During the consultation process the following issues must be considered:

- Undue risk in the event of failure of the drain or sewer.
- Maintain access (manholes/inspection chambers should be retained).
- Provide satisfactory diversionary route to allow sewer to be reconstructed without affecting the building.
- Length of sewer under a building should not exceed 6 m.
- Buildings/extensions should not be constructed over or within 3 m of any drain or sewer more than 3 m deep, or greater than 225 mm in diameter.
- Protect drain or sewer during construction.
- Protect against settlement, e.g. use of minimum 100 mm granular filling.
- Trench should not be excavated below the foundations of any adjoining building. See also Approved Document H1 guidance.

REQUIREMENT H5: SEPARATE SYSTEMS OF DRAINAGE

Any system for discharging water to a sewer which is provided pursuant to paragraph H3 shall be separate from that provided for the conveyance of foul water from the building.

Requirement H5 applies only to a system provided in connection with the erection or extension of a building where it is reasonably practicable for the system to discharge directly or indirectly to a sewer for the separate conveyance of surface water which is:

(a) shown on a map of sewers; or
(b) under construction either by the sewerage undertaker or by some other person, where the sewer is the subject of an agreement to make a declaration of vesting pursuant to Section 104 of the Water Industry Act 1991 (a).

(a) 1991 c. 56

This new requirement brings into the domain of the Building Regulations the need to provide separate drainage to building work. This has previously been covered by local legislation, which has now been repealed. To help minimize the volume of rainwater entering the public foul sewer system, which can overload the capacity of the sewer and cause flooding, provide separate systems of drains and sewers for foul water and rainwater where:

- the rainwater is not contaminated; and
- the drainage is to be connected either directly or indirectly to the public sewer system (separate foul and surface water, or where under construction).

Where the sewerage undertaker has provided or is to provide (i.e. an existing combined system) a separate system of sewers, separate drainage systems will be necessary to comply with Section 106 (Right to communicate with public sewers) of the Water Industry Act 1991.

With respect to contaminated runoff (reference paragraph 3.21 of the Approved Document) pollution must be avoided where materials are stored that could result in contaminated runoff into the surface water drainage system. To connect into the foul water system the consent of the sewerage undertaker should first be obtained.

REQUIREMENT H6: SOILD WASTE STORAGE

1. Adequate provision shall be made for storage of solid waste.
2. Adequate means of access shall be provided:

    (a) for people in the building to the place of storage; and
    (b) from the place of storage to a collection point, where one has been specified by the waste collection authority under Section 46 (household waste) or Section 47 (commercial waste) of the Environmental Protection Act 1990(b), or to a street (where no collection point has been specified).

    (b) 1990 c. 43

The storage and collection of refuse to all buildings should not be prejudicial to health, should have sufficient capacity having regard to frequency of removal, and should be readily accessible for occupants of the building and for removal. The efficacy of a refuse storage system is dependent on its capacity and the ease of removal in relation to the collection service provided by the waste collection authority. Hence early consultation is vital to establish specific authority requirements, noting their powers under Sections 45, 46

and 47 of the Environmental Protection Act 1990. The recycling of household and other waste is an example, which is not covered in Requirement
H6.

Domestic developments

On the basis of a refuse output of 0.09 m³ per dwelling collected weekly,
any house, bungalow or flat in a low-rise development up to four storeys
should have, or have access to, a location where at least two movable
(individual or communal) waste containers can be stored. Combined
capacity to be 0.25 m³ per dwelling, with separate containers for recycled
waste and for waste which cannot be recycled. Where separate storage
areas are provided for each dwelling, an area of 1.2 × 1.2 m² should be
sufficient for the waste containers and space for access. For communal
provisions the waste collection authority must be consulted.

For high-rise/multi-storey domestic developments, with floors above the
fourth floor, a refuse chute should be used feeding a single waste container
(non-recyclable waste). Recycled waste should be stored separately. Where
site constraints dictate or where a satisfactory management arrangement is
in place to convey the waste, storage compounds or rooms can be used.
The use of 'Residents Only' recycling centres is to be encouraged, especially
in large residential developments. Further design guidance is as follows:

- Individual containers should have close-fitting lids.
- Refuse chutes should have a non-absorbent internal surface and close-
  fitting access doors (located at each floor level containing a dwelling),
  and should be ventilated top and bottom. Cross-reference should also
  be made to Approved Documents B and E.
- Any room or chamber used for containers need not be enclosed, but
  if it is, then high- and low-level ventilation should be provided, access
  allowed for filling and emptying (minimum 150 mm clearance around
  containers) and 2 m headroom where communal containers are located.
  Also consider design of access steps/ramps, cross-reference Part K.
- The maximum travel distance for householders to carry refuse to a
  container (including a communal container) or chute should be 30 m.
- Refuse collection vehicle access should be within 25 m of the containers.
- Containers, including dustbins and communal containers, to new
  buildings only should be located so as to avoid collection through a
  building, unless it is a garage, carport or other covered open space and
  have wash down provisions. For conversions it may be necessary for
  the householder to bring the refuse and/or container to the front of
  the property for collection.
- Further detailed guidance with regard to refuse chute systems can be
  found in BS 5906: 1980.

Non-domestic developments

For other building types and where high refuse densities may occur, consultations should take place with the waste collection authority, where proposals to address the following matters should be established:

- volume, nature and storage capacity required based on collection frequency and size/type of container;
- segregation of waste where it can be recycled;
- storage and on-site treatment related to layout and building density;
- location and access, for vehicles and operatives, to storage and treatment areas;
- hygiene arrangements and protection measures against fire hazards.

The **alternative approach** is to refer to BS 5906: 1980 *Code of practice for storage and on-site treatment of solid waste from buildings.* Note that this standard does not include guidance on recycling.

Chapter 11

Approved Document J: Combustion appliances and fuel storage systems

The potential risks associated with the installation of heat-producing appliances in buildings are related to noxious fumes and fire. To combat these problems Requirements J1, J2 and J4 seek to provide for sufficient combustion air, the suitable discharge of combustion products and the protection of the building fabric. Requirements J5, J6 and J7, relating to provision of information, liquid fuel storage and pollution, respectively, have been added by the 2002 edition of the Approved Document. The 2010 Edition added Requirement J3 relating to the provision of carbon monoxide alarms. For work where a building regulation submission would not be necessary, reference should be made to Tables 1.1 and 16.1.

The Approved Document guidance is split into five sections, starting with some guidance and basic provisions that apply generally. The remaining sections deal with solid fuel, gas and oil appliances up to certain specific ratings. Above these ratings specialist advice and/or the alternative approaches may be utilized. It should be noted that electrical heat-producing appliances and portable heaters are not controlled; in addition it is assumed that incinerators burning any fuel are controlled.

REQUIREMENT J1: AIR SUPPLY

Combustion appliances shall be so installed that there is an adequate supply of air to them for combustion, to prevent overheating and for the efficient working of any fire.

Requirement J1 applies only to fixed combustion appliances (including incinerators).

REQUIREMENT J2: DISCHARGE OF PRODUCTS OF COMBUSTION

Combustion appliances shall have adequate provision for the discharge of products of combustion to the outside air.

Requirement J2 applies only to fixed combustion appliances (including incinerators).

REQUIREMENT J3: WARNING OF RELEASE OF CARBON MONOXIDE

Where a combustion appliance is provided, appropriate provision having regard to the design and location of the appliance shall be made to detect and give early warning of the release of carbon monoxide at levels harmful to persons.

Requirement J3 applies only to fixed combustion appliances located in dwellings.

REQUIREMENT J4: PROTECTION OF BUILDING

Combustion appliances and fluepipes shall be so installed, and fireplaces and chimneys shall be so constructed and installed, as to reduce to a reasonable level the risk of people suffering burns or the building catching fire in consequence of their use.

Requirement J4 applies only to fixed combustion appliances (including incinerators).

REQUIREMENT J5: PROVISION OF INFORMATION

Where a hearth, fireplace, flue or chimney is provided or extended, a durable notice containing information on the performance capabilities of the hearth, fireplace, flue or chimney shall be affixed in a suitable place in the building for the purpose of enabling combustion appliances to be safely installed.

REQUIREMENT J6: PROTECTION OF LIQUID FUEL STORAGE SYSTEM

Liquid fuel storage systems and the pipes connecting them to combustion appliances shall be so constructed and separated from buildings and the boundary of the premises as to reduce to a reasonable level the risk of the fuel igniting in the event of fire in adjacent buildings or premises.

Requirement J6 applies only to –

(a) fixed oil storage tanks with capacities greater than 90 litres and connecting pipes; and
(b) fixed liquefied petroleum gas storage installations with capacities greater than 150 litres and connecting pipes;

that are located outside the building and that serve fixed combustion appliances (including incinerators) in the building.

REQUIREMENT J7: PROTECTION AGAINST POLLUTION

Oil storage tanks and the pipes connecting them to combustion appliances shall –

(a) be so constructed and protected as to reduce to a reasonable level the risk of the oil escaping and causing pollution; and
(b) have affixed in a prominent position a durable notice containing information on how to respond to an oil escape so as to reduce to a reasonable level the risk of pollution.

Requirement J7 applies only to fixed oil storage tanks with capacities of 3500 litres or less, and connecting pipes, which are –

(a) located outside the building; and
(b) serve fixed combustion appliances (including incinerators) in a building used wholly or mainly as a private dwelling;

but does not apply to buried systems.

Section 0: General guidance

This section acts as an introduction to the provisions and confirms that the guidance contained within the 2010 Approved Document is aimed mainly at domestic scale installations. Owing to this fact specialist guidance is likely for larger installations; also refer to *CIBSE design guide volume B and practice standards produced by BSI and the Institution of Gas Engineers*. Section 5, offering guidance on Requirement J6 (and J7), relates to heating oil storage installations (up to 3500 litres) and liquefied petroleum gas (LPG) storage installations (up to 1.1 tonne). Section 0 then goes on to provide an explanation of terms used and a number of definitions, for example, a **combustion appliance** (or **appliance**) is the apparatus where fuel is burned to generate heat for space heating, water heating, cooking or other similar purpose, i.e. boilers, warm air heaters, water heaters, fires, stoves and cookers.

Section 1: Provisions that apply generally to combustion installations

Irrespective of the appliance rating, consideration should be given to the following items:

• An appliance should be **room-sealed** (i.e. not reliant on combustion air from within the room), or the room or space in which it is located should be provided with a ventilation opening. This may be via an adjoining room or space that also has the same size opening giving to the external air. Allowance should also be made for cooling air, where required.

- Normally ventilation openings should not be in fire-resisting walls; this may cause problems where the use of a fire shutter and/or fuel cut-off could be used.
- For appliances that are not room-sealed, reference should be made to Approved Document F concerning ventilation, the installation of air extract fans and the guidance provided in this section.
- Open-flued combustions appliances should operate safely where extract fans are provided. In addition to suitable testing, compliance can be shown:

  (a) for gas appliance in kitchen, fan extract rate should not exceed 20 litres/second (72 m³/hour);

  (b) for oil appliance, extract rate should not exceed 40 litres/second (pressure jet burner) or 20 litres/second (vaporising burner);

  (c) for solid fuel appliance avoid extract ventilation. In a kitchen a passive stack may be suitable or otherwise seek specialist advice;

  (d) for commercial and industrial installations seek specialist advice;

  (e) for fans used to extract Radon follow guidance in BRE Good Building Guide 25.

- An appliance should have a balanced flue or low-level flue, or should connect with a flue pipe or chimney that discharges to the external air, unless the appliance can operate without discharging the products of combustion to the outside.
- A flue may only have an opening into it for inspection and cleaning, and the fitting of an explosion door, draught stabilizer or diverter.
- A flue concealed within a void should have minimum 300 mm square inspection hatches within 1.5 m of bends and joints to allow visual inspection.
- Where a flue serves more than one appliance, reference should be made to BS 5410–1: 1997 for oil- and BS 5440–1: 2008 for gas-fired systems; each solid fuel appliance should have its own flue.
- Water condensation should be controlled by insulating flues or use suitable components/allow draining where they serve condensing appliances.
- Chimneys should be constructed with flue liners and masonry suitable for the intended application, i.e. bricks, medium-weight concrete blocks or stone with suitable mortar joints for the masonry and suitably supported and caulked liners. Examples of suitable liners for solid fuel (and other fuel) appliances are those designated T400 N2 D 3 G (as described in BS EN 1443: 2003) clay or concrete flue liners.
- Flueblock chimneys should be constructed with factory-made components suitable for the intended application. Examples for solid fuel (and other fuels) are those designated T400 N2 D 3 G (as described in BS EN 1443: 2003) and blocks lined as above, independently certified as suitable for the purpose.

- A **material change of use** may require improved fire resistance to chimney locations, e.g. within a flat conversion.
- Note that **building work** includes relining (with new rigid or flexible prefabricated components) or a cast in situ liner (that significantly alters the flue's internal dimensions).
- Where a flue in an existing chimney is to be reused it should be tested as outlined in Appendix E of the Approved Document. Chimneys without flues should be lined.
- For factory-made metal chimneys reference is provided in the Approved Document to a range of British Standard guidance. Issues to consider include use of sleeves to allow for thermal expansion; avoidance of joints within construction; isolation from combustible materials and guarding where located in cupboards, etc.
- A notice plate, or similar approved, should be provided for hearths and flues containing the following information:

  (a) location of the hearth/fireplace/flue;
  (b) category of the flue and generic types of appliance that can be used;
  (c) type and size of the flue (or its liner) and the manufacturer's name;
  (d) installation date;

  and be located next to electricity consumer unit/chimney or hearth described/water supply stopcock.

Section 2: Additional provisions for appliances burning solid fuel (including solid biofuel) with a rated output up to 50 kW

Solid fuel appliances are basically open fires, which utilize an open chimney flue, or closed appliances, e.g. boilers, which should use a flue pipe to connect the appliance to a chimney. The guidance contained in the Approved Document can be summarized as follows:

- Air supply, flue sizes and flue outlet locations should be in accordance with Figure 11.1.
- With the exception of a horizontal connection to a chimney, up to 150 mm, flues should be vertical. Where a bend is necessary it should not be at an angle of more than 45° to the vertical.
- A flue pipe should not pass through a roof void.
- Flue pipes may be of cast iron, 3 mm mild steel, 1 mm stainless steel, or vitreous enamelled steel. Spigot and socket joints, where used, should be fitted with sockets upwards.
- Flue pipes should be isolated from the building fabric by an air space of at least 3 × flue diameter or 200 mm of solid non-combustible material, or by using a non-combustible shield, 4 × diameter of flue in

width, where the combustible material is isolated from the shield by 12.5 mm and the flue by 1.5 × diameter.

- Flue pipes should be provided with a debris collection space where the chimney is not directly above the appliance.
- Minimum wall thickness for both brickwork and blockwork is 100 mm, or 200 mm where the wall is between the flue and another compartment, building or dwelling.
- Isolate combustible materials by at least 200 mm from the flue or 40 mm from the chimney face itself (excluding floorboards, skirtings, etc.).
- Tapered gathers, or a suitable canopy, should be provided above open fire places.
- The constructional hearth on which the appliance is located should be of non-combustible material at least 125 mm thick, 840 mm square (for a freestanding appliance) or should project 500 mm in front of a fireplace recess.
- Fireplace recesses should be of solid non-combustible material.
- A carbon monoxide alarm, complying with BS EN 50291, should be installed in a room where a new or replacement fixed solid fuel appliance is to be installed in a dwelling. This should be at least battery powered (working life of alarm) located between 1–3 m horizontally from the appliance, at least 300 mm from wall if ceiling mounted, or 150 mm from ceiling if wall mounted.
- For specific design guidance on fireplace recesses, walls adjacent to hearths and location of appliances reference should be made to the diagrams of the Approved Document Section 2.

The **alternative approach** to the above guidance is to refer to the relevant recommendations of BS 8303: 1994 *Code of practice for installation of domestic heating and cooking appliances burning solid mineral fuels*, BS EN 15287–1: 2007 relating to chimneys.

Section 3: Additional provisions for gas-burning appliances with a rated input up to 70 kW (net)

This section of the Approved Document offers guidance on open-flued and balanced-flued appliances, solid fuel effect fires and cookers all burning gas and up to a rated input of 70 kW. This guidance is described as follows.

In addition to showing compliance with building regulation requirements, consideration must also be given to the Gas Safety (Installation and Use) Regulations. If relevant building work is undertaken by a member of a class of persons approved by the Health and Safety Executive (HSE), that work need not be notified to a Building Control Body. Some of the main requirements of the Gas Safety (Installation and Use) Regulations are as follows:

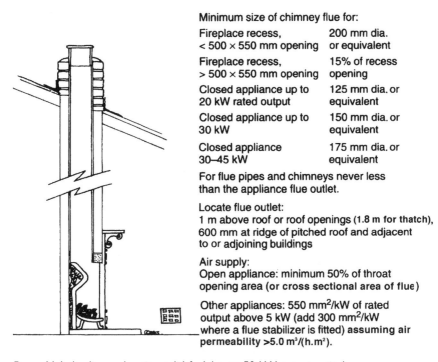

Minimum size of chimney flue for:

Fireplace recess, < 500 × 550 mm opening	200 mm dia. or equivalent
Fireplace recess, > 500 × 550 mm opening	15% of recess opening
Closed appliance up to 20 kW rated output	125 mm dia. or equivalent
Closed appliance up to 30 kW	150 mm dia. or equivalent
Closed appliance 30–45 kW	175 mm dia. or equivalent

For flue pipes and chimneys never less than the appliance flue outlet.

Locate flue outlet:
1 m above roof or roof openings (1.8 m for thatch), 600 mm at ridge of pitched roof and adjacent to or adjoining buildings

Air supply:
Open appliance: minimum 50% of throat opening area (or cross sectional area of flue)

Other appliances: 550 mm²/kW of rated output above 5 kW (add 300 mm²/kW where a flue stabilizer is fitted) assuming air permeability >5.0 m³/(h.m²).

Figure 11.1 Appliances burning solid fuel (up to 50 kW output rating).

- A person must only install gas fittings, appliances and gas storage vessels with the required competence, as verified by the person controlling the work.
- Any appliance installed in a bath or shower room must be of the room-sealed type.
- A gas fire/space heater/water heater of more than 14 kW (gross) heat input must not be installed within sleeping accommodation unless the appliance is room-sealed or equipped with an automatic shutdown device activated before dangerous levels of combustion products are apparent in the room (or cupboard in that room).
- Instantaneous water heaters installed in any room must be as outlined in the bullet item above.
- Precautions must be taken to ensure that all installation pipework, gas fittings, appliances and flues are installed safely, e.g. safe position and effectiveness of the flue and sufficient supply of combustion air.
- Any alterations to maintain compliance with the Regulations.
- LPG storage vessels/fired appliances fitted with automatic ignition devices or pilot lights must not be installed in cellars or basements.

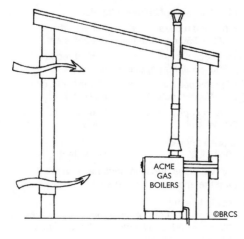

Minimum size of flue:
Gas fire: minimum cross-sectional area of 16 500 mm² for a square flue, 90 mm minimum dimension, and 125 mm diameter for a round flue
Other appliances: minimum cross-sectional area as appliance outlet

Locate flue outlet:
Balanced-flued appliance: to allow free intake of combustion air, dispersal of products of combustion, prevent entry of matter that could block flue, guard terminal if liable to damage or person contact and locate minimum 300 mm from any opening above terminal
Other appliances: located at roof level to allow air flow, minimum 600 mm from any opening and fitted with a flue terminal where dimension less than 170 mm (See also Table Diagram 34 of Approved Document)

Air supply:
Open-flued appliance in a room or space: permanent vent opening minimum 500 mm²/kW of input rating (first 7 kW in older dwellings can be ignored where air permeability more than 5.0 m³/hr/m²)
Cooker: openable window, plus a permanent vent of 10 000 mm² if room less than 5 m³, 5000 mm² if 5–10 m³, if over 10 m³ not required

Figure 11.2 Appliances burning gas (up to 70 kW input rating). Open-flued appliances must not be used in a garage, bathroom or shower room.

The Approved Document guidance can be summarized as follows:

• Air supply, flue sizes and flue outlet locations should be in accordance with Figure 11.2 (and guidance/diagrams contained in the Approved Document).
• Flues should be vertical and avoid horizontal runs. Where a bend is necessary it should not be at an angle of more than 45° to the vertical.
• Installation of simulated coal or wood-effect gas fires should follow manufacturers' instructions or accord to BS 5871: Parts 1, 2 or 3: 2001.
• Flue pipes may be of cast iron, 3 mm mild steel, 1 mm stainless steel, vitreous enamelled steel, other sheet metals (as described in BS EN 1856–2: 2004) or fibre cement. Spigot and socket joints, where used, should be fitted with sockets upwards.

- Flue pipes should be isolated from combustible material by a minimum of 25 mm, or where passing through a wall, floor or roof, be isolated by a non-combustible sleeve with 25 mm air gap (measured from the inner pipe for a double-walled flue). Cross-reference should be made to Approved Document B where flues pass through compartment walls.
- Chimneys should be provided with a debris collection space where the chimney is not lined or constructed of flue blocks.
- The constructional hearth for a back boiler should be non-combustible material at least 125 mm thick, or 25 mm thick on 25 mm high supports. Note that a hearth is not needed where the flame of any appliance is above 225 mm or if the appliance complies with manufacturer's instructions.
- For specific design guidance concerning hearths and appliance locations reference should be made to diagrams of the Approved Document, Section 3.

The **alternative approach** is to refer to the relevant recommendations of one of the following British Standards: BS 5440: Part 1: 2008 and Part 2: 2009, BS 5546: 2000, BS 5864: 2004, BS 5871: Parts 1–3: 2005, BS 6172: 2004, BS 6173: 2001, BS 6798: 2009.

Section 4: Additional provisions for oil-burning appliances with a rated output up to 45 kW

Appliances that burn oil, including liquid bio fuel, operate at varying temperatures, and the guidance that follows reflects this point.

- Air supply, flue sizes and flue outlet locations should be in accordance with Figure 11.3.
- Flues should be vertical and avoid horizontal runs. Where necessary, a bend should not be at an angle of more than 45° to the vertical.
- Where flue gas temperatures, under worst operating conditions, could exceed 250 °C then the provisions for flues and chimneys in Section 2 should be applied. If this temperature is unlikely to be exceeded then follow the provisions outlined within Section 3.
- Where the floor surface temperature under the appliance could exceed 100 °C then a constructional hearth in accordance with Section 2 should be provided. Where this temperature is unlikely to be exceeded then the appliance may stand on a rigid and imperforate sheet of non-combustible material.
- The shielding of the appliance itself should accord with the guidance in Section 3 if the surface temperature of the back or sides of the appliance could exceed 100 °C.

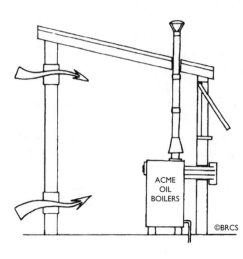

Minimum size of flue pipes not less than the appliance flue outlet

Locate flue outlet:
Balanced-flued appliance: to allow free intake of combustion air, dispersal of products of combustion, prevent entry of matter that could block flue, guard terminal if liable to damage or person contact and locate minimum 600 mm from any opening
Other appliances: with vapourising burner 1 m above roof or roof openings (See also Table Diagram 41 of Approved Document)

Air supply:
Permanent vent minimum 550 mm²/kW of rated output (first 5k W in older dwellings can be ignored where air permeability more than 5.0 m³/hr/m²)

Figure 11.3 Appliances burning oil (up to 45 kW output rating).

A British Standard is referred to as the **alternative approach**, namely BS 5410: *Code of practice for oil firing: Part 1: 1997 Installations up to 45 kW output for space heating and hot water supply purposes.*

Section 5: Provisions for liquid fuel storage and supply

To ensure compliance with Requirements J6 and J7, oil and LPG fuel storage installations, including the pipework connecting them to the combustion appliances, should be reasonably protected from the risk of fire. For oil storage tanks, their ancillary equipment and pipework (dwellings only), precautions must be taken to prevent pollution.

Heating oil storage installations

The Approved Document guidance is limited to oil storage systems with above-ground or semi-buried tanks of 3500 litres capacity or less, used exclusively for heating oil. The relevant recommendations in BS 5410–1: 1997 should be followed, supported by paragraphs 5.4–5.12 from the Approved Document. For tanks with capacities in excess of 3500 litres the local Fire Authority should be consulted and reference can be made to the Control of Pollution (Oil Storage) (England) Regulations 2001.

   To protect against the risk of fire the following criteria should be met:

• Provide ground cover slab (concrete or paving slabs) to prevent over-growing.

- Tank within a building to be in a place of special fire hazard, vented to outside air.
- Adopt 30 min construction where within 1800 mm of a building or provide a fire wall.
- Use a fire wall where less than 760 mm from a boundary.
- Fuel pipework should resist fire and proprietary fire shut-off valve provided.

To protect against the risk of oil pollution, secondary containment, by way of 110% capacity bund, should be provided where one of the following applies:

- total capacity of over 2500 litres; or
- located within 10 m of inland freshwaters or coastal waters; or
- located where spillage could run into an open drain or to a loose-fitting manhole cover; or
- located within 50 m of sources of potable water, such as wells, bore-holes or springs; or
- located where oil spilled from the installation could reach the waters listed above by running across hard ground; or
- located where tank vent pipe outlets cannot be seen from the intended filling point; or
- located within zone 1 (inner protection zone) of an Environment Agency Groundwater Source Protection Zone. Maps available at www.environment-agency.gov.uk/research/library/maps.

Note that a label in a prominent position should offer advice on what to do if an oil spill occurs and the telephone number of the Environment Agency's Emergency Hotline – 0800 80 70 60.

LPG storage installations

These installations are controlled by legislation enforced by the HSE and should be constructed in accordance with an appropriate industry Code of Practice. For installations up to 1.1 tonne capacity, whose tank stands in the open air, the guidance within the Approved Document can be followed.

- Comply with relevant recommendations in the UKLPG Code of practice 1 *Bulk LPG storage at fixed installations Part 1: 2009 and BS 5482–1: 2005.*
- LPG tank should be installed outdoors, and not within an open pit, and be adequately separated from buildings, the boundary and any fixed sources of ignition (refer to Diagram 5.1 of the Approved Document).

- For tanks up to 0.25 tonne capacity, isolate by minimum 2.5 m or 0.3 m if firewall provided. For tanks not exceeding 1.1 tonne capacity, isolate by 3.0 m or 1.5 m if firewall provided. Firewall to be 30 min fire resistant, 60 min if part of the building.
- Drains, gullies and cellar hatches within the separation distances should be protected from gas entry.

Where LPG is stored as a set of cylinders the following should be considered:

- Cylinders to stand upright, on a firm level base (e.g. 50 mm concrete) secured by straps or chains against a wall in a well-ventilated position at ground level.
- Be readily accessible, reasonably protected from physical damage, not obstructing exit routes from the building.
- Located so that cylinder valves will be minimum 1 m horizontally and 300 mm vertically from openings or heat sources (e.g. flue terminals and tumble-dryer vents) and minimum 2 m horizontally from drains without traps, unsealed gullies or cellar hatches (unless 250 mm high intervening wall provided).

The Approved Document concludes with:

Appendix A – Checklist for checking and testing of hearths, fireplaces, flues and chimneys – copy reproduced opposite.
Appendix B – Opening areas of large or unusual fireplaces.
Appendix C – Example calculation of the ventilation requirements of a gas-fired appliance.
Appendix D – Example calculation of the ventilation requirements of an oil-fired appliance.
Appendix E – Methods of checking compliant with Requirement J2.
Appendix F – Assessing air permeability of older dwellings in relation to permanent ventilation requirements.
Appendix G – European chimney designations.
Appendix H – Addresses, a list of 35 British Standards and other useful/ referred to publications.

CHECKLIST
Hearths, fireplaces, flues and chimneys

This checklist can help you to ensure hearths, fireplaces, flues and chimneys are satisfactory. If you have been directly engaged, copies should also be offered to the client and to the Building Control Body to show what you have done to comply with the requirements of Part J. If you are a subcontractor, a copy should be offered to the main contractor.

1. Building address, where work has been carried out ....
   ....
   ....

2. Identification of hearth, fireplace, chimney or flue.

3. Firing capability: solid fuel / gas / oil / all.

4. Intended type of appliance. State type or make. If open fire give finished fireplace opening dimensions.

5. Ventilation provisions for the appliance: State type and area of permanently open air vents.

6. Chimney or flue construction
   a) State the type or make and whether new or existing.
   b) Internal flue size (and equivalent height, where calculated – natural draught gas appliances only).
   c) If clay or concrete flue liners used confirm they are correctly jointed with socket uppermost and state jointing materials used.
   d) If an existing chimney has been refurnished with a new liner, type or make of liner fitted.
   e) Details of flue outlet terminal and diagram reference.
      Outlet Detail:
      Complies with:
   f) Number and angle of bends.
   g) Provision for cleaning and recommended frequency.

7. Hearth. Form of construction. New or existing?

8. Inspection and testing after completion
   Tests carried out by:
   Tests (Appx E In AD J) and results
   Flue   visual
   Inspection   sweeping
                                        coring ball
                                             smoke
          Appliance (where included) spillage

I/We the undersigned confirm that the above details are correct. In my opinion, these works comply with the relevant requirements in Part J of Schedule 1 to the Building Regulations.
Print name and title.. Profession.....................................
Capacity.. Tel no.......................................
Address...Postcode................
Signed.. Date...
Registered membership of ... (e.g. GasSafe, OFTEC, HETAS, NACE, NACS)...........................

Chapter 12

Approved Document K: Protection from falling, collision and impact

People, including children and the elderly, should be able to use stairs safely and be protected from the risk of falling from one floor level to others below. The degree of safety provisions necessary depends on the purpose group of the building, the number of users, their familiarity with the building and the extent of access made available to the building. The guidance contained in Approved Document K reflects its close relationship with Approved Document B, relating to means of escape, and Approved Document M, covering access to buildings for the disabled, where overlapping guidance is now included in Approved Document K. An example of this is the omission of minimum stair and ramp widths for escape routes, although the minimum widths from Approved Document M are now included. The 2013 edition has also taken the opportunity to incorporate the guidance from Approved Document N: Glazing – safety in relation to impact, opening and cleaning. This now gives rise to the use of ten sections to cover the extended list of requirements.

The main requirements of the Workplace (Health, Safety and Welfare) Regulations 1992, relating to building design, are covered by the Building Regulations. Compliance with Requirements K1–K4, K5.1–K5.4 and K6, in accordance with Section 23(3) of the Health and Safety at Work, etc. Act 1974, would prevent the serving of an improvement notice under the Workplace Regulations, unless the requirements of the regulations are more onerous themselves.

REQUIREMENT K1: STAIRS, LADDERS AND RAMPS

Stairs, ladders and ramps shall be so designed, constructed and installed as to be safe for people moving between different levels in or about the building. Requirement Kl applies to stairs, ladders and ramps which form part of the building.

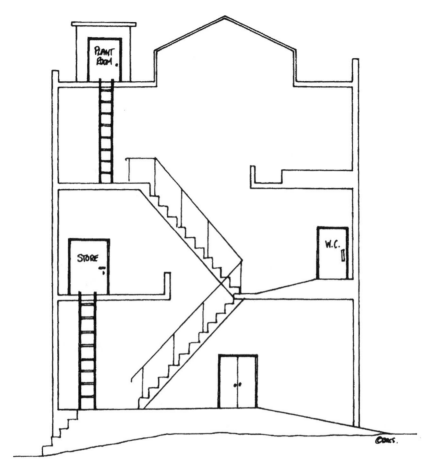

Figure 12.1 Application of stair and ramp provisions.

The wording of the requirement and the limitations of application, as illustrated in Figure 12.1, are relatively clear. Steps or ramps on land surrounding a building are not controlled, although an entrance ramp or flight of steps would be controlled, as they can be regarded as part of the building. Differences in level less than 600 mm in a dwelling, and up to two risers (or 380 mm, if not part of a stair) in other buildings are not controlled as far as guarding is concerned. For mixed-use developments (including dwellings) the requirements for non-domestic use can be applied to the shared parts of the building.

Section 1 of the Approved Document starts with the design of stairs; ramps are covered in Section 2. Key terms are now listed in Appendix A, although the significant ones are as follows:

Common stair – serves more than one dwelling.

Flight – the part of the stair or ramp between landings formed by a continuous series of steps or a continuous ramp.

General access stair – for all users on a day-to-day basis as a normal route between levels.

Helical stair – a helix round a central void.

Pitch – the slope/inclination of the line connecting all tread nosings of a flight.

Private stair – serves only one dwelling.

Spiral stair – a helix round a central column.

Utility stair – for escape or maintenance purposes only, not used on a day-to-day basis.

Section 1: Stairs and ladders

The first design criterion to consider is the steepness of stairs. Three categories of stairs are specified with corresponding maximum rise and minimum going limitations (the rise is measured between the top surfaces of each tread, and the going dimension is measured from nosing to nosing; all risers and goings in a flight should be the same):

- **private stair**, maximum 220 mm rise (minimum 150 mm) and minimum 220 mm going (maximum 300 mm), and a maximum pitch of 42°;
- **utility stair**, maximum 190 mm rise (minimum 150 mm) and minimum 250 mm going (maximum 400 mm);
- **general access stair**, maximum 170 mm rise (minimum 150 mm) and minimum 250 mm going (maximum 400 mm). (For schools 150 mm rise and 280 mm going is preferred.)

The relationship, 2 × rise + going = 550–700 mm can also be utilized for design purposes.

For gangways serving seated spectators, the maximum pitch should not exceed 35°. In addition, the following should also be followed:

- align rows so gangways remain the same width;
- provide traverse gangways to allow side access/exit;
- avoid traverse gangways and radial gangways crossing;
- stepped tiers should use equal risers 100 mm minimum and 190 mm maximum;
- a pitch exceeding 25°, uninterrupted by gangways, should use a maximum of 40 steps;
- provide a minimum 1100 mm landing at exit from a stepped gangway;
- provide a handrail for stepped side gangways;
- seatway and nearest step should be the same level in stepped tiers; and

- gangway minimum width should be 1100 mm, or 900 mm for up to 50 persons.

The other provisions contained in Section 1, and to which building they apply, can be summarized as follows:

- Open riser stairs should *only* be used in dwellings, with a minimum tread overlap of 16 mm and no gap that would allow the passage of a 100 mm diameter sphere (the approximate size of a child's head).
- Distinguish all nosings (except for dwellings) with a material that will contrast visually, minimum 50–65 mm wide (tread) and 30–55 mm (riser) for common stair and minimum 55 mm wide (tread and riser) for all other stairs. Maximum step nosing projection should be 25 mm.
- Access routes between levels, including flights of stairs, should have minimum headroom of 2 m. For a stair serving a loft conversion this may be reduced to 1.9 m at the centre of the stair (1.8 m to the side of the stair).
- Where a soffit under a stair (which does not serve dwellings) is less than 2 m, provide guarding and low-level cane detection or an equivalent barrier.
- For minimum stair widths, where used for means of escape, reference should be made to Approved Document B: Fire safety – Volume 2. Where a stair is not used for means of escape, then the width should be 1200 mm between walls, strings or upstands (1000 mm between handrails); and be divided into minimum 1000 mm widths with handrails where over 2 m wide.
- For dwellings no control exists on the width of a stair unless a steeply sloping site requires a stair within the entrance storey, where a width of 900 mm should be used.
- For buildings other than dwellings stair flights should be limited to 16 risers for a **utility stair** and 12 risers (16 for small premises) in the case of a **general access stair**. Single steps should not be used.
- For stairs that incorporate consecutive flights with more than 36 risers, a change of direction of at least 30° should be introduced by way of a landing, the dimensions of which should be at least the width of the stair.
- A clear, level and unobstructed landing should be provided at both the top and bottom of each flight, the dimensions of which should be at least the width of the stair. Two concessions to this are allowable, in that any door may swing across a landing located at the *bottom of the flight* (*dwelling only*) and any cupboard or duct doors may swing across a landing at the top or bottom of the flight. For both concessions a minimum 400 mm landing depth, to the full width of the stair, should be retained.

- Landings formed by the ground may be at a gradient not steeper than 1:60.

Section 1 isolates particular provisions for special stairs, which can be listed as follows:

- The going for steps with tapered treads should be measured at the centre for stair widths not exceeding 1000 mm, or 270 mm in from each side for a width 1000 mm or over; minimum tread width 50 mm.
- Consecutive tapered treads should all have the same going. Tapered treads combined with a straight flight should have the same going as the straight flight.
- For **spiral** and **helical stairs** reference should be made to BS 5395: Part 2: 1984 (as amended) *Code of practice for the design of helical and spiral stairs.*
- An **alternating tread stair** may only be used in a loft conversion to give access to one habitable room, with a bathroom and/or WC (but not the only WC). Space is saved by alternate cutaway treads which retain the required rise and going dimensions. Treads should have slip-resistant surfaces, and handrails are necessary to both sides of the stair.
- Where no other option is available, consideration can be given to the use of a fixed ladder to serve just one habitable room. It should have handrails both sides and not be retractable if used for means of escape.
- For industrial buildings specific design guidance may be necessary, where reference should then be made to BS 5395: Part 3: 1985 (as amended) *Code of practice for the design of industrial stairs, permanent ladders and walkways,* or BS 4211: 2005 (as amended) *Specification for permanently fixed ladders.*

Guidance is given on handrails and guarding to stairs:

- For a stair width not exceeding 1000 mm a handrail should be provided to one side, and to both sides for a width 1000 mm or over (and within a dwelling entrance storey on a steeply sloping site).
- Handrail heights should be 900–1000 mm to all buildings, measured from the pitch line or floor up to the top of the handrail (600 mm for any second/lower handrail provided).

Except for dwellings and flats provided with a lift, all handrails should be:

- continuous on each side of the flights *and* landings;
- extend 300 mm onto the landing, but not to project into the access route;
- terminate to avoid clothes getting caught;

- contrast visually with the background;
- be slip-resistant and not hot or cold to the touch in extreme hot/cold temperatures – use of relatively low thermal conductivity metals may be considered, e.g. stainless steel;
- adopt a suitable profile, circular (32–50 mm diameter) or non-circular (50 mm wide and 39 mm deep, with rounded edges).
- Stairs, including flights and landings, should be guarded at the sides as outlined in Table 12.1. Differences in level of less than 600 mm in a dwelling, and up to two risers in other buildings, are not controlled.
- For stairs likely to be used by children under five years the guarding should have no gap that would allow the passage of a 100 mm diameter sphere. Children should also not be able to climb the guarding.
- Finally in Section 1, less demanding provisions apply for access to areas of maintenance where the frequency of visits is taken into account. Access at least once a month could utilize domestic stairs or a ladder to BS 5395: Part 3. Where required less frequently a portable ladder could be used, where temporary means of access are covered by the Construction (Design and Management) Regulations 2007.

Table 12.1 Guarding design

Building category	For the location specified use guarding height of		
	800 mm	*900 mm*	*1100 mm*
Single-family dwellings		Stairs, landings, ramps, edges of internal floors	External balconies and roof edges
Factories and warehouses		Stairs and ramps	Landings and floor edges
Residential, office, institutional, retail, educational and public buildings		For flights	All other locations
Assembly	530 mm in front of fixed seating	For flights	All other locations
All buildings (except roof windows in loft conversions)	Opening windows and glazing at changes of level		

Section 2: Ramps

Since ramps, forming part of the building, are often used to allow disabled people access and circulation within the building the relevant guidance from Approved Document M has now been incorporated. For the design of ramps, cross-reference should also be made to Section 1, with specific regard to landings and guarding. The remaining provisions for an approach steeper than 1:20, i.e. a ramp, can be summarized as follows:

- Be readily apparent or clearly sign-posted (not needed for dwellings).
- Not be steeper than 1:20 (flight 10 m, maximum rise 500 mm), 1:15 (flight 5 m, maximum rise 333 mm) or 1:12 (flight 2 m, maximum rise 166 mm). No flight should have a going greater than 10 m, or a rise of more than 500 mm.
- Ramp to incorporate suitable complementary steps where the rise of the ramp exceeds 300 mm (2 × 150 mm steps) (not for dwellings).
- Should provide a slip-resistant surface, contrasting colour to landings, but with similar frictional characteristics (not for dwellings).
- Provide a minimum 100 mm kerb, with visual contrast, to open-sided flights and landings (in addition to any guarding required) (not for dwellings).
- Where a soffit under a ramp (which does not serve dwellings) is less than 2 m, provide guarding and low-level cane detection or an equivalent barrier.
- Provide landings at the foot and head of the ramp, with an unobstructed length of at least 1200 mm. Intermediate landings to have an unobstructed length of at least 1500 mm (1800 mm square if a wheelchair user is unable to see from one end of the ramp to the other). All landings should be level or a maximum length gradient of 1:60. See also guidance in Section 1.
- For minimum ramp widths, where used for means of escape, reference should be made to Approved Document B: Fire safety – Volume 2. Otherwise to be minimum 1500 mm wide.
- Provide handrails to accord with Section 1. A ramp not more than 600 mm high serving dwellings need not have handrails.
- Provide guarding for ramps and landings as Section 1.

REQUIREMENT K2: PROTECTION FROM FALLING

(a) Any stairs, ramps, floors and balconies and any roof to which people have access,
(b) and any light well, basement area or similar sunken area connected to a building

shall be provided with barriers where it is necessary to protect people in or about the building from falling.

Requirement K2(a) applies only to stairs and ramps that form part of the building.

Section 3: Protection from falling

Section 3 deals with the requirements for both pedestrian guarding and vehicle barriers, the application of which is illustrated in Figure 12.2.

The first factor to consider for pedestrian guarding is where it should reasonably be provided so as to prevent people from being injured by falling from a height above 600 mm in a dwelling, and from falling in the first place in any other building above a height of two risers (or 380 mm, if not part of a stair). The locations that should be considered are: the edge of any floor (including an opening window), gallery, balcony, roof (including rooflights and other openings); places to which people have access (unless for the purpose of maintenance or repair only); any light well, basement or similar sunken area next to or within a building. Guarding need not be provided on ramps used only for vehicle access or to loading bays where it would present an obstruction to normal use of the area.

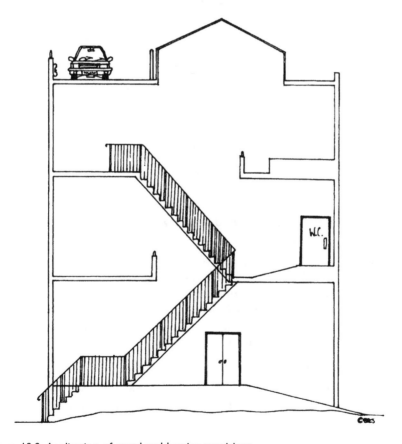

Figure 12.2 Application of guard and barrier provisions.

The design of the guarding may take the form of a wall, parapet, balustrade, screen or similar obstruction, which should be at a minimum height as indicated in Table 12.1. From a structural point of view the guarding should resist a horizontal force applied at the top of the guarding: reference should be made to BS EN 1991-1-1 and PD 6688-1-1 (and BS 6180: 2011 for infill panels).

Where the guarding utilizes glazing, reference should be made to Section 5. Finally, in the design of pedestrian guarding, consideration should be given to whether the building is likely to be used by children under five years. If so, no gap in the guarding should allow the passage of a 100 mm diameter sphere, and children should not be able to climb the guarding.

Less demanding provisions apply for access to areas of maintenance where the frequency of visits is taken into account. Access at least once a month could utilize domestic stairs, or where less frequent access is required temporary guarding and/or warning notices could be used covered by the Construction (Design and Management) Regulations 2007.

REQUIREMENT K3: VEHICLE BARRIERS AND LOADING BAYS

1. Vehicle ramps and any levels in a building to which vehicles have access, shall be provided with barriers where it is necessary to protect people in or about the building.
2. Vehicle loading bays shall be constructed in such a way, or be provided with such features, as may be necessary to protect people in them from collision with vehicles.

Section 4: Vehicle barriers and loading bays

Turning to the need for vehicle barriers, these should be provided to the edges of any floor, roof or ramp to which vehicles have access at or above ground or upper floor levels. Barrier design may take the form of a wall, parapet, balustrade or similar obstruction, which should be at a minimum height, for any building, of 375 mm to a floor or roof edge and 610 mm at a ramp edge. The barrier should resist a horizontal force applied at the top of the barrier, where reference should be made to BS EN 1991-1-1 and PD 6688-1-1.

Loading bays should have at least one exit point, i.e. steps, or two exit points for wide loading bays used by more than three vehicles. A refuge could also be used as an alternative for one exit point in order to avoid people being crushed.

REQUIREMENT K4: PROTECTION AGAINST IMPACT WITH
GLAZING

Glazing, with which people are likely to come into contact whilst
moving in or about the building, shall:

(a) if broken on impact, break in a way that is unlikely to cause
    injury; or
(b) resist impact without breaking; or
(c) be shielded or protected from impact.

Section 5: Protection against impact with glazing

Requirement K4, and K5.2, seek to control the installation of glazing in
critical locations in the building. A number of accidents have resulted in
serious injuries and fatalities where people, both young and old, have fallen
or walked into areas of glazing.

It is important to note that new glazing forming part of the erection,
extension or material alteration of a building is defined as **building work**,
and would thus need to comply with these two requirements. The installation
of replacement glazing, on the other hand, is not controlled, although
consumer protection legislation may apply. For an exempt extension, e.g.
a conservatory not exceeding 30 m², a submission under the Building Regulations would not be necessary, although the glazing used should comply
with Requirement K4, and K5.2 if applicable.

To limit the risk of people in or about the building sustaining cutting or
piercing injuries, certain critical locations are identified. These are illustrated
by the shaded areas in Figure 12.3, and recognize the potential areas of
impact at door positions and at low level, where children are particularly
at risk.

To show compliance with Requirement K4, one of four options for glazing
in critical locations may be chosen.

Safe breakage of a glazing material is defined in BS EN 12600 and BS
6206: 1981 (as amended) *Specification for impact performance requirements
for flat safety glass and safety plastics for use in buildings.* For panes in
doors, and door side panels more than 900 mm wide, a Class 2 (BS EN
12600) or Class B (BS 6206) material should be used as a minimum. For
other critical locations a Class 3 (BS EN 12600) or Class C (BS 6206)
material may be used, noting that Class 1 (BS EN 12600) or Class A (BS
6206) is the highest. This classification, together with the material code,
BS number and product name or trademark, should be permanently marked
on the material.

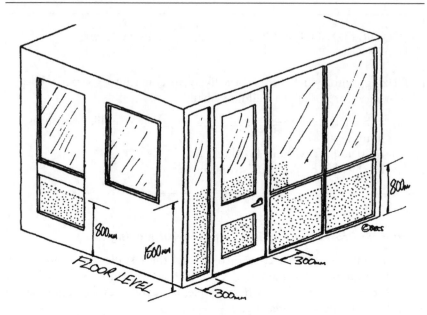

Figure 12.3 Critical locations in internal and external walls.

Robustness is an inherent quality of a range of glazing materials, i.e. polycarbonate sheet or glass blocks. Annealed glass gains strength with thickness and can be used in large areas, e.g. shopfronts, although it should not be used in doors. As a guide, 10 mm annealed glass can be used for a maximum pane size of 2.25 m², whereas no limits apply to a thickness of 15 mm.

Glazing in small panes relates to the use of a number of small, isolated panes and includes traditional leaded lights or copper lights. The pane should not have a width exceeding 250 mm nor have an area greater than 0.5 m². Annealed glass should be a minimum of 6 mm thick (4 mm in leaded or copper lights).

Permanent screen protection calls for the provision of a robust screen, which should prevent a 75 mm sphere from coming into contact with the glazing, where the glazing itself would not need to satisfy Requirement K4. If the screen also serves as guarding then it should be unclimbable.

REQUIREMENT K5.1: PROTECTION FROM COLLISION WITH OPEN WINDOWS, ETC.

Provision shall be made to prevent people moving in or about the building from colliding with open windows, skylights or ventilators.

Requirement K5.1 does not apply to dwellings.

Section 6: Protection from collision with open windows, etc.

The requirement basically applies the guidance originally contained in Approved Document M, although it does not apply to dwellings.

To avoid collision with the projecting parts of windows, skylights and ventilators, both internally and externally, adopt one of the options described:

- restrict opening to 100 mm; or
- keep projection at least 2 m above ground or floor level; or
- if below 2 m, provide a feature, barrier or rail, 1100 mm high; or
- use tactile differences, cobbles or a planting strip.
- For spaces used only infrequently and for maintenance, reliance could be placed on clear marking of the projection.

REQUIREMENT K5.2: MANIFESTATION OF GLAZING

Transparent glazing, with which people are likely to come into contact while moving in or about the building, shall incorporate features that make it apparent.

Requirement K5.2 does not apply to dwellings.

Section 7: Manifestation of glazing

Critical locations with respect to this requirement include large, uninterrupted areas of transparent glazing, e.g. internal office screens or external walls to shops, factories, public buildings, etc. These elements, especially if located on floors that are at the same level, may give the impression that access is possible from one part of the building to another without interruption. To avoid impact in such circumstances glazing should have permanent **manifestation** (in the form of lines, patterns or logos) or some other means to indicate the existence of these large areas of glazing. Figure 12.4 illustrates where manifestation should be provided and where it would not be required owing to the design of doors or the size of glazing panels. Reference should also be made to the provisions of Part M.

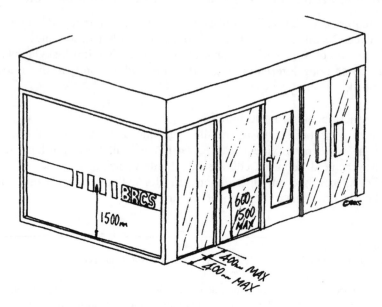

Figure 12.4 Methods to indicate large areas of transparent glazing.

REQUIREMENT K5.3: SAFE OPENING AND CLOSING OF
WINDOWS, ETC.

Windows, skylights and ventilators which can be opened by people in or
about the building shall be so constructed or equipped that they may be
opened, closed or adjusted safely.

Requirement K5.3 does not apply to dwellings.

Section 8: Safe opening and closing of windows, etc.

To comply with this requirement, some options are described. Guidance
for the operation of windows to sleeping accommodation is also provided
in Section 4 of Approved Document M.

• Height of controls should not be more than 1.9 m where unobstructed.
• If obstructed, the height of the controls should be lowered; for example,
  1.7 m where an obstruction is 900 mm high and 600 mm deep.

- Where controls cannot be suitably positioned, then consideration will need to be given to remote operation (manual or electrical). This may be appropriate for disabled people.
- Where there is a danger of falling through a window (above ground level), then suitable opening limiters or appropriate guarding should be installed (cross-reference Section 3).

REQUIREMENT K5.4: SAFE ACCESS FOR CLEANING WINDOWS, ETC.

Provision shall be made for any windows, skylights or any transparent or translucent walls, ceilings or roofs to be safely accessible for cleaning.

Requirement K5.4 does not apply to:

(a) dwellings, or
(b) any transparent or translucent elements whose surfaces are not intended to be cleaned.

Section 9: Safe access for cleaning windows, etc.

A series of options to allow safe cleaning of both sides of the glazing where someone could fall more than 2 m are described:

- Use a suitably designed window that allows the outside surface to be cleaned from the inside. Diagram 8 of the Approved Document gives safe reach dimensions and further reference can be made to BS 8213: Part 1 *Windows, doors and rooflights.*
- Use of a portable ladder, not more than 9 m long, placed on an adequate area of firm ground, 75° pitch and provided with suitable tying/fixing point if over 6 m long.
- Use of minimum 400 mm wide walkways with 1100 mm high guarding or anchorage for sliding safety harness.
- Use of access equipment (with safety harnesses) – for example: suspended cradle, travelling ladders or a 'cherry picker'.
- Use of suitable anchorage points for safety harnesses/abseiling hooks.
- As a last resort, allow sufficient space for the erection of a temporary tower scaffold.

REQUIREMENT K6: PROTECTION AGAINST IMPACT FROM
AND TRAPPING BY DOORS
1.   Provision shall be made to prevent any door or gate:
(a) that slides or opens upwards, from falling onto any person; and
(b) that is powered, from trapping any person.
2.   Provision shall be made for powered doors and gates to be opened in
the event of a power failure.
3.   Provision shall be made to ensure a clear view of the space on either
side of a swing door or gate.

Requirement K5 does not apply to:

(a) dwellings, or
(b) any door or gate which is part of a lift.

Section 10: Protection against impact from and trapping by doors

A series of options to prevent the opening and closing of doors and gates presenting a safety hazard are described:

• Provide a visibility zone of glazing to cover a height of 900–1500 mm above the floor level, unless low enough to see over, i.e. 900 mm.
• Provide effective stops at ends of sliding tracks and retaining rail, should the suspension system fail.
• Provide an effective device to prevent an upward opening door or gate (e.g. roller shutter) from falling.
• Provide safety features to power-operated doors and gates, i.e. pressure sensitive door edge/stop switch/opening override in case of power failure.

People, especially those with sight impairments, using access routes up to and around the outside of the building should be protected at outward-opening door locations (excluding fire escapes). Projections over 100 mm should have suitable guarding, barriers or low-level cane detection.

Chapter 13

Approved Document L: Conservation of fuel and power

The 2013 edition of Part L continues to be a tool for government to reduce carbon emissions from buildings, the energy from which accounts for nearly half of all carbon dioxide emissions in the UK. Part L of the Building Regulations is also used to comply with the EU Directive on the Energy Performance of Buildings (EPBD). The EPBD required the government to develop and implement a National Calculation Method (NCM) with a view to being able to set targets for maximum carbon dioxide emissions from buildings and to facilitate the ability to issue energy performance certificates based on the results. To achieve this, the government developed the Standard Assessment Procedure for Energy Rating of Dwellings (SAP) as the NCM for dwellings and the Simplified Building Energy Model (SBEM) for buildings other than dwellings.

The Welsh government has produced separate Approved Documents which show how to comply with the Regulations in Wales from 31 July 2014. It is likely that the guidance in Wales and England will continue to diverge over the coming years, with the Welsh government keen to take a faster path to zero-carbon buildings.

The technical requirements applicable in England and Wales are, however, broadly similar. This text concentrates on the guidance in the Approved Documents for England, with reference being made to the main differences in the Welsh Approved Documents. The 2013 Part L requires a 6% carbon dioxide saving relative to Part L 2010 for dwellings (8% in Wales), and a 9% improvement for new non-domestic buildings (20% in Wales). These improvements are averaged across the mix of building and dwelling types, e.g. offices, retail units, mid-terraced houses, detached houses, etc.

To aid understanding of the text some key terms and abbreviations used in the guidance are reproduced below:

Air permeability is the physical property used to measure air tightness of the building fabric. It is defined as air leakage rate per hour, per envelope area at the test reference pressure differential across the building envelope of 50 Pa (50 N/m^2). The envelope area of the building, or measured

---

REQUIREMENT L1: CONSERVATION OF FUEL AND POWER

Reasonable provision shall be made for the conservation of fuel and power in buildings by:

(a) limiting heat gains and losses:
    (i) through thermal elements and other parts of the building fabric;
    (ii) from pipes, ducts and vessels used for space heating, space cooling and hot water services;

(b) providing fixed building services which:
    (i) are energy efficient;
    (ii) have effective controls; and
    (iii) are commissioned by testing and adjusting as necessary to ensure they use no more fuel and power than is reasonable in the circumstances

---

part of the building, is the total area of all floors, walls and ceilings bordering the internal volume subject to the test. This includes walls and floors below external ground level. Overall internal dimensions are used to calculate this area, and no subtractions are made for the area of the junctions of internal walls, floors and ceilings with exterior walls, floors and ceilings.

BER is the Building CO_2 Emission Rate expressed as $kgCO_2/(m^2/year)$.

Consequential improvement means those energy efficiency improvements required by Regulation 28.

DER is the Dwelling CO_2 Emission Rate expressed as $kgCO_2/(m^2/year)$.

Design air permeability is the value of air permeability selected by the designer for use in the calculation of the BER or DER.

Display lighting means lighting intended to highlight displays of exhibits or merchandise, or lighting used in spaces for public leisure and entertainment such as dance halls, auditoria, conference halls, restaurants and cinemas.

Display window means an area of glazing, including glazed doors, intended for the display of products or services on offer within the building, positioned:

(a) at the external perimeter of the building;

(b) at an access level and immediately adjacent to a pedestrian thoroughfare.

There should be no permanent workspace within one glazing height of the perimeter. Glazing more than 3 m above such an access level should not be considered part of a display window except:

(a) where the products on display require a greater height of glazing;

(b) in existing buildings, when replacing display windows that already extend to a greater height;

(c) in cases of building work involving changes to the facade and glazing requiring planning consent, where planners should have discretion to require a greater height of glazing, e.g. to fit in with surrounding buildings or to match the character of the existing facade.

Dwelling means a self-contained unit designed to accommodate a single household.

Dwelling type is a means of allocating each dwelling on a development to a particular group to provide the basis for the air-testing regime. To be classed as the same type, dwellings have to:

- be of the same generic form, e.g. detached, end terrace;
- be of the same number of storeys;
- be of the same design air permeability;
- have similar adjacency to unheated spaces, e.g. integral garages;
- have the same principal construction details;
- have a similar (±1) number of significant penetrations, i.e. windows, doors, fans, flues;
- have envelope areas that do not differ by more than 10%.

Energy efficiency requirements means the requirements of Regulations 23, 25A, 25B, 26, 26A, 26B, 28, 29 and 40 and Part L of Schedule 1 to the Building Regulations [Regulations 25B and 26C in Wales].

Fixed building services means any part of, or any controls associated with:

(a) fixed internal or external lighting systems, but does not include emergency escape lighting or specialist process lighting; or

(b) fixed systems for heating, hot water service, air conditioning or mechanical ventilation.

High-usage entrance door means a door to an entrance primarily for the use of people that is expected to experience large traffic volumes, and where robustness and/or powered operation is the primary performance requirement. To qualify as a high-usage entrance door, the door should be equipped with automatic closers and, except where operational requirements preclude, be protected by a lobby.

Principal works means the work necessary to achieve the client's purposes in extending the building and/or increasing the installed capacity of any fixed building services. The value of the principal works is the basis for determining a reasonable provision of consequential improvements.

Renovation in relation to a thermal element means the provision of a new layer in the thermal element or the replacement of an existing layer, but excludes decorative finishes, and 'renovate' shall be construed accordingly.

Room for residential purposes means a room, or a suite of rooms, that is not a dwelling house or a flat and that is used by one or more persons to live and sleep. This includes a room in a hostel, a hotel, a boarding house, a hall of residence or a residential home, whether or not the room is separated from or arranged in a cluster group with other rooms, but does not include a room in a hospital, or other similar establishment, used for patient accommodation. For the purposes of this definition, a 'cluster' is a group of rooms for residential purposes; that is, separated from the rest of the building in which it is situated by a door that is designed to be locked; and not designed to be occupied by a single household.

Simple payback means the amount of time it will take to recover the initial investment through energy savings, and is calculated by dividing the marginal additional cost of implementing an energy efficiency measure, by the value of the annual energy savings achieved by that measure, taking no account of VAT.

(a) The marginal additional cost is the additional cost (materials and labour) of incorporating, e.g. additional insulation, not the whole cost of the work.

(b) The cost of implementing the measure should be based on prices current at the date the proposals are made known to the Building Control Body and be confirmed in a report signed by a suitably qualified person.

(c) The annual energy savings should be estimated using SAP 2012 or the SBEM.

(d) The energy prices current at the time of the Building Regulation application should be used where evaluating the annual energy savings.

TER is the target CO_2 emission rate expressed as $kgCO_2/(m^2/year)$.

TFEE is the Target Fabric Energy Efficiency rate expressed as $kWh/(m^2/year)$.

Total useful floor area is the total area of all enclosed spaces measured to the internal face of the external walls, that is to say it is the gross floor area as measured in accordance with the guidance issued to surveyors by the RICS. In this convention:

(a) the area of sloping surfaces such as staircases, galleries, raked auditoria and tiered terraces should be taken as their area on plan; and

(b) areas that are not enclosed such as open floors, covered ways and balconies are excluded.

Approved Document LIA: Conservation of fuel and power in new dwellings

Section I: The requirements

Section 1 lists five criteria that should be met in order to comply with the energy efficiency requirements for newly erected dwellings:

Criterion 1: the calculated rate of carbon dioxide emissions from the dwelling, the Dwelling Emission Rate (DER), must not be greater than the Target Emission Rate (TER). Also the calculated Dwelling Fabric Energy Efficiency (DFEE) rate, must not be greater than the Target Fabric Energy Efficiency (TFEE) rate.

Criterion 2: the performance of the individual fabric elements and the fixed building services of the building should achieve reasonable overall standards of energy efficiency.

Criterion 3: the dwelling should have appropriate passive control measures to limit the effect of heat gains on indoor temperatures in summer irrespective of whether the dwelling has mechanical cooling.

Criterion 4: the performance of the dwelling, as built, should be consistent with the DER and the DFEE rate.

Criterion 5: the necessary provisions for the energy efficient operation of the dwelling should be put in place.

Section 2: Design standards

The **Target CO_2 Emission Rate** (TER) is one of two minimum energy performance requirements for new dwellings in England and is expressed in terms of the mass of carbon dioxide in units of kilograms per square metre (kg/m^2) of floor area per year, emitted as a result of the provision of heating, hot water, ventilation and internal fixed lighting for a standardized household, when assessed using SAP 2012.

The target is calculated by first calculating the carbon dioxide emissions from a notional dwelling, of the same size and shape as the actual dwelling, which is constructed according to the reference values set out in Appendix R of SAP 2012 and summarized in Table 13.4. This will give the carbon dioxide emissions arising from the provision of heating and hot water and the use of internal fixed lighting. The TER is then calculated using the following formula:

$$TER_{2013} = C_H \times FF + C_{PF} + C_L$$

where

C_H = carbon dioxide arising from heating and hot water;
C_L = carbon dioxide emitted from internal fixed lighting;
C_{PF} = carbon dioxide arising from pumps and fans; and
FF = fuel factor

Fuel factor depends on fuel type, as shown in Table 13.1. The fuel factor used to calculate the TER should be based on the following fuel:

- where all the heating appliances are served by the same fuel, the fuel used in those appliances;
- where a dwelling has more than one heating appliance and these are served by different fuels, the fuel used for the TER calculation should

Table 13.1 Fuel factors

Heating fuel	Fuel factor
Mains gas	1.00
LPG	1.06
Oil	1.17
Grid electricity (for direct acting, storage and electric heat pump systems)	1.55
Solid mineral fuel	1.35
Any fuel with a CO_2 emission factor less than that of gas	1.00
Solid multi-fuel	1.00
B3OK	1.00

be mains gas if any of the heating appliances are fired by mains gas, or if not, the fuel used in the main heating system;
- where a dwelling is served by a community heating scheme, mains gas if used for any purpose in the scheme, or if not, the fuel that provides the most heat.

Where a building contains more than one dwelling, an average TER can be calculated for all the dwellings in the building. In such cases, the average TER is the floor-area-weighted average of all the individual TERs, and is calculated according to the following formula:

{(TER 1 × floor area 1) + (TER 2 × floor area 2) +

(TER 3 × floor area 3) + . . .)} ÷ {(floor area 1 +

floor area 2 + floor area 3) + . . .}

The **Target Fabric Energy Efficiency** (TFEE) rate is the second and new minimum energy performance requirement for a new dwelling erected in England. It is calculated using SAP 2012 by determining the fabric energy efficiency from a notional dwelling and multiplying the result by 1.15. Where there are multiple dwellings an average TFEE rate can be calculated in the same way as for the average TER described above, replacing TER with TFEE in the formula. [Note that in Wales there is a similar but separate requirement, Regulation 25C, to achieve minimum fabric performance values.]

Criterion 1: Achieving the TER and TFEE rate

The proposed **Dwelling CO_2 Emission Rate** (DER) and the **Dwelling Fabric Energy Efficiency** (DFEE) rate must be no worse than their respective targets, i.e. the TER and TFEE rate. The design stage DER and DFEE rate calculations must be submitted to the Building Control Body at least one day prior to commencement on site. The as-built DER and DFEE rate must be submitted to the Building Control Body not later than five days after completion of the dwelling and must be based on the dwelling as constructed, incorporating:

(a) any changes to the design stage specifications; and
(b) the assessed air permeability, which is determined as follows:

- the air test for the dwelling where it has been tested;
- where the dwelling has not been tested but is of the same type as a tested dwelling on the same development, the assessed air permeability is the average test result for the dwelling type plus 2 $m^3/hr/m^2$ at 50 Pa;

- on small developments where the developer has opted to avoid testing, the assessed air permeability is 15 m³/hr/m² at 50 Pa.

When calculating the DER it may be assumed that the central heating system is supplemented by a secondary heating system. This must be selected based on the following:

- Where a secondary heating appliance is fitted, the efficiency of the actual appliance with its appropriate fuel should be used in the calculation of the DER.
- Where a chimney or flue is provided but no appliance is actually installed, then the presence of the following appliances shall be assumed:

  1. If a gas point is located adjacent to the hearth, a decorative fuel effect fire open to the chimney or flue with an efficiency of 20%.
  2. If there is no gas point, then an open fire in a grate with an efficiency of 37% burning multi-fuel, unless the dwelling is in a smoke-control area, when the fuel should be taken as smokeless solid mineral fuel.
  3. Otherwise assume the secondary heater has the same efficiency and fuel as the main heating system.

With respect to lighting, the DER is calculated assuming the proportion of low energy lighting actually installed.

Where a building contains more than one dwelling, Regulation 26 is achieved if either every individual dwelling has a DER that is no greater than its corresponding TER, or if the average DER is no greater than the average TER. The average DER is the floor-area-weighted average of all the individual DERs and is calculated in the same way as the average TER described earlier. Similarly, Regulation 26A is met using the same approach for the TFEE rate calculation.

Achieving the TER and TFEE rate

As long as the dwelling satisfies the limits on design flexibility in Criterion 2, the designer is free to achieve the TER and TFEE rate as they please.

Regulation 25A requires an analysis of the feasibility of using high-efficiency alternative systems to be carried out for each dwelling, groups of dwellings or common types of dwelling proposed within the same area. The person undertaking the work must give notice to the Building Control Body before work commences that the analysis has been done and the results must be available for inspection should they be requested.

Special considerations

COMMON AREAS IN BUILDINGS WITH MULTIPLE DWELLINGS

Common areas of buildings containing multiple dwellings should be separated from the dwellings with fabric elements meeting the required U-value if they are unheated. If the common area is heated, the guidance in L2A for buildings other than dwellings should be followed.

CONSERVATORIES AND PORCHES

If a conservatory or porch is built as part of the new dwelling with thermal separation from the main dwelling, and the heating system is not extended into the conservatory or porch, the guidance in Approved Document L1B should be followed in respect of the construction of the conservatory itself. If the conservatory or porch is integral with the dwelling, or the heating system is extended into the area, the space should be included as part of the new dwelling when checking against the five compliance criteria.

SWIMMING POOL BASINS

Where a swimming pool is constructed as part of a new dwelling, the pool basin should achieve a U-value no worse than 0.25 W/m^2 K. For the purpose of calculating the DER the dwelling should be assessed as if the pool were not there, i.e. the area of the basin should be replaced with the same area of floor with the same U-values as the pool surround.

Criterion 2: Limits on design flexibility

Design limits apply to the building envelope; each element of the building fabric should have U-values within the limiting standards shown in Table 13.2.

Table 13.2 Limiting U-value standards [Welsh figures in brackets]

Element	Area-weighted average U-value (W/m^2 K)	
Wall	0.3	[0.21]
Party wall	0.2	[0.20]
Floor	0.25	[0.18]
Roof	0.2	[0.15]
Windows, roof windows, rooflights, doors, curtain walling, and pedestrian doors	2.0	[1.60]

Notes
U-values to be calculated following BR443 'Conventions for U-value Calculations' BRE 2006.
U-value of a window, roof window or rooflight, or door unit can be taken as the value for either the standard configuration set out in BR443; or the particular size and configuration of the actual unit; or the smaller of the two standard windows defined in BS EN 14351-1

The limit for the *design air permeability* is 10 m³/(h m²) at 50 Pa.

System efficiencies. All fixed building services need to comply with the 2013 edition of the *Domestic Building Services Compliance Guide*. Where an innovative product or system is proposed that is not covered within the guide the system should be as efficient as a comparable system within the guide.

Criterion 3: Limiting the effects of solar gains in summer

Limiting internal temperature rise due to solar gains is achieved by a combination of window size and orientation, solar protection through shading and other solar control measures. SAP 2012 Appendix P contains a procedure to check whether solar gains are excessive. Reasonable provision is achieved if the SAP assessment indicates that the dwelling will not have a high risk of high internal temperatures.

Heat losses and gains from circulation pipes

Pipe heat losses and gains should be minimized by insulating to the minimum standards set out in the *Domestic Building Services Compliance Guide*, which now requires domestic hot water primary circulation pipes to be insulated throughout their length. It is also recommended that primary space heating pipes passing through common areas of flats are insulated in addition to the domestic hot water pipes.

Section 3: Quality of construction and commissioning

Criterion 4: Building performance consistent with DER and DFEE rate

The as-built performance of the dwelling must meet the limiting standards and efficiencies, and in addition, the DER and DFEE rate for the dwelling must reflect the as-built details. Criteria 4 and 5 are concerned with ensuring that this is the case.

Party walls and other thermal bypasses. Dwelling party walls are considered as heat loss walls, with the notional dwelling having zero heat loss through its party wall. In calculating the DER for the as-built dwelling the U-value that will be applied will depend on the construction of the wall, as illustrated in Table 13.3.

Thermal Bridges. There are four options available to show that sufficient provision has been made to prevent reasonably avoidable thermal bridges in the insulation layer:

- Use DCLG Approved Construction Details or other details formally recognized by DCLG.

Table 13.3 U-values for party walls

Party wall construction	U-value (W/m² K)
Solid	0.0
Unfilled cavity with no effective edge sealing [not permitted in Wales}	0.50
Unfilled cavity with effective sealing around all exposed edges and in line with insulating layers in abutting elements	0.2
Fully filled cavity with effective sealing around all exposed edges and in line with insulating layers in abutting elements	0.0

- Use details calculated by a competent person in accordance with BRE Report BR 497.
- Use linear thermal transmittance values from the 'default' column of table K1 in SAP 2012 in the DER and DFEE rate calculations.
- Use the default y value of 0.15 W/m² K in the DER and DFEE rate calculations.

Pressure testing is required by virtue of Regulation 43 to demonstrate that the specified air permeability has been achieved. On each development an air pressure test should be carried out on the lesser of three units of each dwelling type or 50% of that type.

The requirement will have been met if the measured air permeability is not worse than 10 m³/(h m²) at 50 Pa, and the calculated DER and DFEE rate using the measured air permeability, is not worse than the TER and TFEE rate. If satisfactory performance is not achieved, then remedial measures should be carried out on the dwelling and a new test carried out until the dwelling achieves the above criteria. In addition to the remedial work on a dwelling that failed the initial test, one additional dwelling of the same dwelling type should be tested.

As an alternative approach to specific pressure testing on development sites where no more than two dwellings are to be erected, it is acceptable to demonstrate that during the preceding 12-month period a dwelling of the same dwelling type constructed by the same builder had been pressure tested and had achieved the specified air permeability. It is also possible to avoid the need for any pressure testing by using a value of 15 m³/(h m²) for the air permeability at 50 Pa when calculating the DER.

Commissioning of heating and hot water systems

The heating and hot water system(s) should be commissioned so that, at completion, the system(s) and their controls are left in the intended working order and can operate efficiently for the purposes of the conservation of fuel and power. In order to demonstrate that the heating and hot water systems have been adequately commissioned, the procedure in Regulation 44 should be followed as set out in the *Domestic Heating Compliance Guide* and the *Domestic Ventilation Compliance Guide*. The notice required by Regulation 44 should include a declaration signed by a suitably qualified person that the manufacturer's commissioning procedures have been completed satisfactorily.

Section 4: Providing information

Criterion 5: Provisions for the energy efficient operation of the dwelling

- The owner of the building should be provided with sufficient information about the building, the fixed building services and their maintenance

"....AND HERE ARE ALL YOUR HANDOVER DOCUMENTS SO YOU CAN OPERATE YOUR NEW ECO HOUSE IN AN ENERGY EFFICIENT MANNER....!"

requirements so that the building can be operated in such a manner as to use no more fuel and power than is reasonable.

- Suitable instructions and manuals should be provided in a way that householders would understand, to include floor plans showing the main heating and ventilation components, with an explanation of how to operate, control and maintain the building services.
- The data used to calculate the DER and TER should be provided along with the Energy Performance Certificate and recommendations report.

Section 5: Model designs

The TER and TFEE rates are based on a notional dwelling of the same size and shape as the actual dwelling constructed to a concurrent specification. Table 13.4 shows a summary of the notional dwelling specification; if the actual dwelling is built to this specification it will meet the requirements of Criterion 1. Some builders may prefer to adopt model design packages rather than to engage in design for themselves. These model packages of fabric U-values, boiler seasonal efficiencies, window opening allowances, etc., would show a compliant overall performance if sufficiently robust.

Appendices

The appendices in L1A contain an explanation of the key terms and abbreviations (Appendix A), confirmation that the document is only used for newly erected dwellings and not those created by a material change of use (Appendix B) and guidance on reporting evidence of compliance (Appendix C). Lists of the documents and standards referred to are in Appendices D and E.

Approved Document L I B: Conservation of fuel and power in existing dwellings

Sections I, 2 and 3: Introduction, the requirements and general guidance

The first three sections provide background information on the use of the Approved Document, sources of further information and definitions. The relevant Regulations are also reproduced. Information is provided on competent person self-certification schemes and confirmation is given that document L1B should be used for all works relating to dwellings other than the erection of new dwellings.

Table 13.4 Summary of concurrent notional dwelling specification

Element or System	Values
Opening areas (windows and doors)	Same as actual dwelling up to a maximum proportion of 25% of total floor area
External walls (including opaque elements of curtain walls)	0.18 W/m^2 K
Party walls	0.0 W/m^2 K
Floor	0.13 W/m^2 K
Roof	0.13 W/m^2 K
Windows, roof windows, glazed rooflights and glazed doors	1.4 (Whole window U-Value) g-value= 0.63
Opaque doors	1.0 W/m^2 K
Semi glazed doors	1.2 W/m^2 K
Air tightness	5.0 m^3/hr/m^2
Linear thermal transmittance	Standardised psi values − See SAP Appendix R, except use of y = 0.05 W/m^2 K if the default value of y = 0.15 W/m^2 K is used in the actual dwelling
Ventilation type	Natural (with extract fans)
Air conditioning	None
Heating system	Mains gas If combi boiler in actual dwelling, combi boiler; otherwise regular boiler Radiators Room sealed Fan flue SEDBUK 2009 89.5% efficient
Controls	Time and temperature zone control Weather compensation Modulating boiler with interlock
Hot water storage system	Heated by boiler (regular or combi as above) If cylinder specified in actual dwelling, volume of cylinder in actual dwelling. If combi boiler, no cylinder. Otherwise 150 litres. Located in heated space. Thermostat controlled Separate time control for space and water heating
Primary pipework	Fully insulated
Hot water cylinder loss factor (if specified)	Declared loss factor equal or better than 0.85 × (0.2 + 0.051V2/3) kWh/day
Secondary space heating	None
Low energy lighting	100% Low energy Lighting
Thermal mass parameter	Medium (TMP = 250)

Section 4: Guidance relating to building work and Section 5: Guidance on thermal elements

The extension of a dwelling

First, the **controlled fittings** should meet the standards shown in Table 13.5.

New thermal elements should meet the standards in Table 13.6, be reasonably airtight and have continuity of insulation, which can be demonstrated by adopting accredited construction details available from www.planningportal.gov.uk.

When working on **existing fabric elements** that are to become **thermal elements**, the U-value should be improved if the existing U-value is worse than the given threshold value shown in Table 13.7, subject to the improvement being economically, technically and functionally feasible. [Improvement required irrespective of existing U-value in Wales, subject to its feasibility.]

The area of **allowable windows, roof windows and doors** in extensions is limited such that it should not exceed the sum of 25% of the floor area of the extension, plus the area of any windows or doors that, as a result

Table 13.5 Standards for controlled fittings W/m² K

Fitting	Dwellings
Window, roof window and rooflight	U-value = 1.6 W/m² K or Window Energy Rating Band C
Doors	1.8 W/m² K or Door Set Energy Rating Band E [1.6 in Wales]

Table 13.6 Standards for thermal elements W/m² K [Figures for dwellings or buildings of domestic character, followed by 'other' buildings in Wales only]

Element	Standard (W/m² K)	
Wall	0.28	[0.21, 0.26]
Pitched roof – insulation at ceiling level	0.16	[0.15, 0.15]
Pitched roof – insulation at rafter level	0.18	[0.15, 0.18]
Flat roof or roof with integral insulation	0.18	[0.15, 0.18]
Floors	0.22	[0.18, 0.22]
Swimming pool basin	0.25	[0.25, 0.25]

Notes
1 'Roof' includes the roof parts of dormer windows and 'wall' refers to the wall parts (cheeks) of dormer windows.
2 A lesser provision may be appropriate where meeting such a standard would result in a reduction of more than 5% in the internal floor area of the room bounded by the wall.
3 A lesser provision may be appropriate where meeting such a standard would create significant problems in relation to adjoining floor levels. The U-value of the floor of an extension can be calculated using the exposed perimeter and floor area of the whole enlarged building.

Table 13.7 Standards for upgrading retained thermal elements

Element	Threshold value (W/m² K)	Improved value (W/m² K)
Cavity wall	0.70	0.55
Other wall type	0.70	0.30
Pitched roof – insulation at ceiling level	0.35	0.16
Pitched roof – insulation at rafter level	0.35	0.18
Flat roof or roof with integral insulation	0.35	0.18
Floors	0.70	0.25

Notes
Improvements are subject to a test of technical, functional and economic feasibility. The upgrade should be done to the best technical and functional standard that can be achieved within a simple payback period of 15 years.

If a cavity wall is unsuitable for cavity insulation, treat it as 'other wall type'.

Where an existing thermal element is being renovated it need only be improved if work is being done to more than 25% of its surface area.

The flat roof improved value may be increased to 0.2 W/m² K where there are technical limitations, e.g. upstand heights.

of the extension works, no longer exist or are no longer exposed. Occasionally more glazing may be preferred in order to meet BS 8206-2: 1992 *Lighting for buildings: Code of practice for daylighting*. This would require the agreement of the Building Control Body.

Where a **fixed building service** is provided or extended as part of constructing the extension it should be installed and commissioned following guidance in the *Domestic Building Services Compliance Guide 2013*. When replacing appliances the new appliance should not be significantly less efficient than the appliance being replaced and a replacement renewable energy generator should have an electrical output of not less than the original generator. Where possible, consideration should be given to connecting to local heat networks.

Two **alternative methods** of showing compliance are described that allow greater design flexibility. The first is to show that the area-weighted U-value for all elements in the extension is no greater than that of an extension of the same size and shape that complies with the U-value standards in Table 13.6 and the opening area described above.

Where even greater design flexibility is required, SAP 2012 can be used to show that the calculated carbon dioxide emission rate from the dwelling with its proposed extension is no greater than for the dwelling plus a notional extension built to the compliant standards. The area-weighted average U-value of each element type should be no worse than the limiting standards set out in Table 13.6. The data in SAP 2012 Appendix S can be used to estimate the performance of the elements of the existing building where these are unknown. Any improvements that are proposed to the existing dwelling to help achieve compliance should be implemented to a standard that is no worse than set out in the Approved Document.

Where the extension is substantially glazed or a **conservatory**, which is not exempt, adopt the following:

- Effective thermal separation from the heated area in the existing dwelling. The walls, doors and windows between the dwelling and the extension should be insulated and draught stripped to at least the same extent as in the existing dwelling.
- Independent temperature and on/off controls to any heating system.
- Glazed elements and thermal elements should meet the U-values in Tables 13.5 and 13.6, respectively.

Material change of use

When carrying out a material change of use the Approved Document gives the following guidance:

- Where controlled services or fittings are being provided or extended they should meet the guidance as described above for extensions.
- Where the work involves the provision, retention or renovation of a thermal element it should be provided or improved in accordance with the appropriate tables above, depending on whether the element is new, or being retained.
- Any existing window (including roof window or rooflight) or door that separates a conditioned space from an unconditioned space or the external environment, and that has a U-value that is worse than 3.3 W/m^2 K, should be replaced to achieve the U-value for fittings listed in Table 13.5.

To provide more design flexibility SAP 2012 can be used to demonstrate that the total carbon dioxide emissions from all the dwellings in the building as it will become are no greater than if each dwelling had been improved following the above guidance.

Material alteration

When carrying out a material alteration reasonable provision is considered to be as follows:

- Where the work involves the provision, retention or renovation of a thermal element it should be provided or improved in accordance with the appropriate table above, depending on whether the element is new, or retained.

- When providing controlled fittings, limit glazing area to 25% of the floor area of the dwelling unless compensatory measures are included elsewhere in the work and meet the U-value in Table 13.5.
- Where controlled services are being provided or extended they should meet the guidance as described above for extensions.

Section 6: Consequential improvements to energy performance

This section confirms that **consequential improvements** apply to existing dwellings with a floor area exceeding 1000 m² if it is extended, has an initial installation of a fixed building service or an **increase in the installed capacity of a fixed building service**. This excludes renewable energy generators. Further guidance is provided in Approved Document L2B. [In Wales consequential improvements are required for all dwelling extensions and conversions such as loft and garage conversions. The extent of improvements required depends on whether the extension is less or more than 10 m².]

Section 7: Providing information

As for new dwellings, sufficient information must be given to the building owner on completion of works to enable them to operate and maintain their property and the building services in an energy efficient manner.

Appendix A gives additional guidance on what might constitute reasonable provision in most circumstances when renovating thermal elements, while acknowledging the fact that there may be technical and functional limitations. References for further advice are listed together with a table giving examples of considerations when undertaking typical proposed works to various constructions.

Approved Document L2A: Conservation of fuel and power in new buildings other than dwellings

Section 1: The requirements

The route to showing compliance is based on meeting essentially the **same five criteria** described earlier for new dwellings, except that in England there is no requirement to meet a target energy efficiency rate for buildings other than dwellings. [Note that in Wales Regulation 26A has been introduced, requiring that the calculated Building Primary Energy Consumption (BPEC) must not be greater than the Target Primary Energy Consumption (TPEC). This is concerned with limiting the primary power consumed by the building's fixed building services and must be proved in order to show compliance with Criterion 1 in Wales.]

Section 2: Design standards

The **Target CO_2 Emission Rate** (TER) is the minimum energy performance requirement specified in Regulation 25 and is the mass of carbon dioxide emitted per year per square metre of the total useful floor area of the building ($kg/m^2/year$). The TER must be calculated using one of the calculation tools included in the methodology for calculating the energy performance of buildings approved by the Secretary of State, i.e. the Simplified Building Energy Model (SBEM) or other approved software tools.

The TER is established by using approved software to calculate the CO_2 emission rate from a notional building of the same size and shape as the actual building, but with properties specified in the 2013 NCM modelling guide, the key components of which are listed in Table 13.9.

Criterion 1: Achieving the TER

The **Building CO_2 Emission Rate** (BER) is calculated using the same calculation tool as is used to establish the TER, and to meet Criterion 1 the BER must be no greater than the TER. Regulation 27 requires the BER, TER and a list of the specifications used in the calculations to be submitted to the Building Control Body at least the day before commencement of work on site. The as-built BER and TER have to be submitted to Building Control within five days of completion of the work and must include confirmation that the building has been constructed in accordance with the design stage specifications, and if not, a list of the changes to those specifications.

For the purpose of producing the BER, the CO_2 emission factors to use are specified in SAP 2012.

When systems are capable of being fired by more than one fuel, then:

- for biomass-fired systems supplemented by an alternative appliance, the carbon dioxide emission factor should be based on a weighted average for the two fuels based on the anticipated usage of the fuels;
- where the same appliance is capable of burning both biofuel and fossil fuel, the carbon dioxide emission factor for dual-fuel appliances should be used, except where the building is in a smoke-control area, when the anthracite figure should be used;
- in all other cases, the fuel with the highest carbon dioxide emission factor should be used.

If thermal energy is supplied from a district or community heating or cooling system, emission factors should be determined by considering the particular details of the scheme. Calculations should take account of the annual average performance of the whole system and the BER submission should be accompanied by a report, signed by a suitably qualified person, detailing how the emission factors have been derived.

Certain management features, such as providing automatic monitoring and targeting of energy use, with alarms for out-of-range values, offer improved energy efficiency and help achieve the TER. Where such features are provided, the BER can be reduced by an adjustment factor. Reference should be made to Table 1 of the Approved Document.

Regulation 25A, requiring the person carrying out the work to consider high-efficiency alternative systems described above for dwellings, also applies to new buildings other than dwellings.

Special considerations

Modular and portable buildings with a planned time of use of more than two years. There is recognition that these buildings have special considerations because they tend to have less embodied energy through reuse and because they are often used in 'distress' situations where quick accommodation is needed. To reflect this, the guidance is less onerous than for other buildings. Where at least 70% of the external envelope of the building is created from sub-assemblies manufactured before 6 April 2014, Table 2 of the Approved Document sets out adjustment factors that are applied to the TER, depending on the date of manufacture. Where the building will be used at more than one location for less than two years, the BER and TER are calculated when the building is first manufactured; when relocated confirmation should be given to Building Control that it still meets the original 'generic' module and that the activities assumed in the original calculations remain representative.

Due to these buildings often requiring electrical resistance heating, a further concession is allowed such that it is reasonable to provide energy efficiency measures that are 15% better than using fossil fuel heating by assuming a 77% gas boiler in the BER and TER calculations.

Shell and core developments. Where a building is built as a shell for future fit-out a design-stage BER/TER calculation is required which should assume efficiencies for services not yet installed and a specification should be given to the Building Control Body clearly identifying these. When the shell is fitted out for the first time the BER/TER calculation should be submitted to Building Control based on the installed building services and shell as constructed.

Industrial sites, workshops and non-residential agricultural buildings other than those with low energy demand. Where such buildings are not exempt from the energy efficiency requirements by virtue of having low energy demand but where the NCM would not adequately account for the building use without leading to technical risk, reasonable provision is to provide fixed building services that satisfy the standards in Approved Document L2B.

Criterion 2: Limits on design flexibility

The design limits seek to ensure that heat gains and losses are reasonable and also that reasonably efficient fixed building services with effective control are installed.

Table 13.8 shows the limiting building fabric parameters allowable in new buildings other than dwellings with the U-value being the area-weighted average value for all elements of that type.

In buildings with high internal gains, a higher area-weighted average U-value for the glazing may be an appropriate way of reducing the BER; for example, if the energy used by cooling plant would be reduced. If this case can be made, then the average U-value for windows can be relaxed to be no worse than 2.7 W/m^2 K. The limit for individual glazing elements should not be exceeded unless there are exceptional circumstances, such as constraints imposed by planning authorities. Display windows are not required to meet the U-value for glazing but their impact on the BER calculation must be included in the calculations.

To achieve reasonable standards of energy efficiency in use, the following control features are recommended for **heating, ventilation and air conditioning system controls:**

- The systems should be subdivided into separate control zones to correspond to each area of the building that has a significantly different solar exposure, or pattern, or type of use.
- Each separate control zone should be capable of independent timing and temperature control and, where appropriate, ventilation and air recirculation rate.
- The provision of the service should respond to the requirements of the space it serves. Heating and cooling should be controlled so as not to operate simultaneously.
- Central plant should only operate as and when the zone systems require it, with the default position being off.

Table 13.8 Limiting fabric parameters

Roof	0.25 W/m^2 K
Wall	0.35 W/m^2 K
Floor	0.25 W/m^2 K
Swimming pool basin	0.25 Wm^2 K
Windows, roof windows, rooflights, curtain walling and pedestrian doors	2.2 W/m^2 K
Vehicle access and similar large doors	1.5 W/m^2 K
High-usage entrance doors	3.5 W/m^2 K
Roof ventilators (inc. smoke vents)	3.5 W/m^2 K
Air permeability	10 $m^3/hr/m^2$ at 50 Pa

Energy metering systems should be used to enable at least 90% of the estimated annual energy consumption of each fuel to be assigned to the various end-use categories (heating, lighting, etc.). Refer to *TM39 Building energy metering*, 2006. The performance of any low and zero carbon (LZC) system is to be separately monitored, and in buildings with a total useful floor area greater than 1000 m², provide automatic meter reading and data collection facilities.

For the **heating and hot water service system(s), cooling plant, air handling plant, insulation of pipes, ducts and vessels,** the guidance in the *Non-domestic Heating, Cooling and Ventilation Compliance Guide*, 2013 should be followed with respect to plant efficiencies, controls and insulation.

Criterion 3: Limiting the effects of solar gains in summer

Any space within a building that is **occupiable** or mechanically cooled must be checked to demonstrate that the solar gains through the glazing aggregated between April and September inclusive are no greater than would occur through one of the reference glazing systems below. The NCM database defines how a particular space is daylit, e.g. offices are side lit. The reference cases are as follows:

- Side lit is an east-facing facade with 1.0 m high, full-width glazing having 10% **framing factor** and normal **solar transmittance (g-value)** of 0.68.
- Top lit with a zone height of less than 6 m is a horizontal roof with 10% glazing having a framing factor of 25% and a g-value of 0.68.
- Top lit with an average zone height greater than 6 m has a horizontal roof with 20% glazing having a framing factor of 15% and a g-value of 0.46.

For this part of the Building Regulations an **occupiable space** is a space intended to be occupied by the same person for a substantial part of the day.

Section 3: Quality of construction and commissioning

Criterion 4: Building performance consistent with BER

This section of the Approved Document is concerned with ensuring that the as-built construction meets the requirements of Part L and the following guidance is given.

To demonstrate **continuity of insulation** over the whole building envelope the guidance is similar to that described for new dwellings above. That is, the builder can choose to adopt a quality assured and accredited system, employ a competent person to calculate thermal bridges, use unaccredited details or a combination of each. The BER calculation will be adjusted to reflect the choices made, with the highest rate of thermal transmittance assumed for unaccredited details.

Regulation 43 requires new buildings to be **pressure tested**. The test should follow the procedure given in *Measuring air permeability of building envelopes*, Air Tightness Testing and Measurement Association, with the exceptions being the following:

- Buildings of less than 500 m² total useful floor area where a pressure test can be avoided if a design permeability of (15 m³/h m²) at 50 Pa is used in the calculation of the BER.
- Factory-made modular buildings of less than 500 m² floor area, with a planned time of use of more than two years at more than one location, where no site construction is required other than linkages between standard modules using standard link details. In this case it is necessary to confirm to Building Control that the as-built configuration conforms to the standard module and link details that have been tested on other sites. The average test result must be better than the design air permeability used in the BER calculation by at least 1.0 m³/h m².
- Large extensions, assessed as if they were new buildings, where sealing off the extension from the existing building is impractical; where it is agreed that testing is impractical, the extension should be treated as a large, complex building, with the guidance in the following paragraph then applying.
- Large complex buildings, where owing to building size or complexity, it may be impractical to carry out pressure testing of the whole building. In such cases a suitably qualified person, i.e. a competent person, could be appointed to undertake a detailed programme of design development, component testing and site supervision to give confidence that a continuous air barrier will be achieved; a minimum air permeability of 5.0 m³/(h m²) at 50 Pa should therefore be used in the calculation of the BER.
- Compartmentalized buildings, where buildings are divided into self-contained units with no internal connections in which case pressure tests could be carried out on a representative area of the building.

To show pressure test compliance the measured air permeability should be no worse than 10 m³/(h m²) at 50 Pa and the BER calculated using the measured air permeability is not worse than the TER. If a test failure occurs then remedial measures should be carried out on the building and new tests carried out until the building achieves the criteria.

The building services systems should be **commissioned** so that at completion, the system(s) and their controls are left in working order and can operate efficiently for the purposes of the conservation of fuel and power. Under Regulation 44 a notice is required to be given to the Local Authority confirming that the fixed building services have been commissioned in accordance with the procedures in the following guides:

- *CIBSE Commissioning code M on commissioning management*, CIBSE;
- *DW/143 A practical guide to ductwork leakage testing*, B&ES.

The notice should include a declaration confirming that:

- a commissioning plan has been followed so that every system has been inspected and commissioned in an appropriate sequence and to a reasonable standard; and
- the results of tests confirm that the performance is reasonably in accordance with the actual building designs, including written commentaries where excursions are proposed to be accepted.

Ductwork leakage testing should be carried out in accordance with the procedures set out in *DW/143* on systems served by fans with a design flow rate greater than 1 m³/s and on sections of ductwork with a pressure class where *DW/143* recommends testing.

If a ductwork system fails to meet the leakage standard, remedial work should be carried out as necessary to achieve satisfactory performance in retests, and further ductwork sections should be tested as set out in *DW/143*.

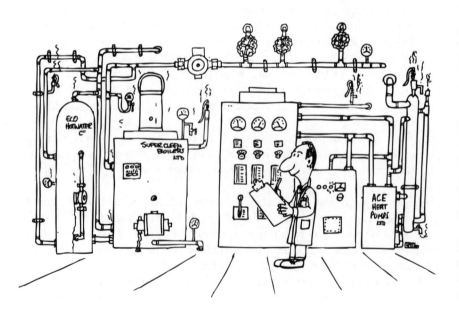

Section 4: Providing information

Criterion 5: Provisions for energy efficient operation of the building

The guidance in L2A is similar to that given for Criterion 5 above for dwellings. A way of showing compliance would be to follow the guidance in *CIBSE TM 31 Building Log Book Toolkit*, CIBSE, 2006. The log book should include the data used to calculate the TER and the BER, together with the recommendations report generated with the Energy Performance Certificate.

Section 5: Model designs

The guidance here is as for model designs in L1A in that the builder may develop a package of U-values, services efficiencies, air permeability rate, etc. that is sufficiently robust to achieve compliance. To achieve the TER the actual building must meet or exceed the specification of the concurrent notional building. The full details of the notional building are given in the *NCM Modelling Guide 2013*, but a summary is given in Table 13.9.

Appendices

Appendix A lists the key terms and abbreviations and Appendix B confirms the types of work covered by L2A, which are:

- the construction of new buildings other than dwellings;
- fit-out works where this is included as part of the construction of the building, or is the first fit out of a shell and core development;
- the construction of extensions to existing buildings that are not dwellings, where the total useful floor area of the extension is greater than 100 m² and greater than 25% of the total useful floor area of the existing building;
- when constructing a building that contains dwellings the parts of the building that are not a dwelling such as heated common areas and, in the case of mixed-use developments, the commercial or retail space.

Appendix C lists the buildings that are exempt from the energy efficiency requirements:

- Places of worship, although this only relates to parts of the building used for, or directly linked with, formal public worship. Ancillary areas such as offices, meeting halls, etc., are not exempt.
- Temporary buildings with a planned time of use of two years or less.

Table 13.9 Summary of concurrent notional building specification

Element	Side lit or unlit (where HVAC specification is heating only)	Side lit or unlit (where HVAC specification includes cooling)	Top lit
Roof U-value (W/m² K)	0.18	0.18	0.18
Wall U-value (W/m² K)	0.26	0.26	0.26
Floor U-value (W/m² K)	0.22	0.22	0.22
Window U-value (W/m² K)	1.6 (10% FF)	1.6 (10% FF)	N/A
G-Value (%)	40%	40%	N/A
Light transmittance (%)	71%	71%	N/A
Roof light U-value (W/m² K)	N/A	N/A	1.8 (15% FF)
G-Value (%)	N/A	N/A	55%
Light transmittance (%)	N/A	N/A	60%
Air-permeability (m³/m²/hour) Gross internal area less than or equal to 250 m²	5	5	7
Air-permeability (m³/m²/hour) Gross internal area greater than 250 m² and less than 3500 m²	3	3	7
Air-permeability (m³/m²/hour) Gross internal area greater than or equal to 3500 m² and less than 10 000 m²	3	3	5
Air-permeability (m³/m²/hour) Gross internal area greater than or equal to 10 000 m²	3	3	3
Lighting luminaire (lm/circuit watt)	60	60	60
Occupancy control (Yes/No)	Yes	Yes	Yes
Daylight control (Yes/No)	Yes	Yes	Yes
Maintenance factor	0.8	0.8	0.8
Constant illuminance control	No	No	No
Heating efficiency (heating and hot water)	91%	91%	91%
Central ventilation SFP (W/l/s)	1.8	1.8	1.8
Terminal unit SFP (W/l/s)	0.3	0.3	0.3
Cooling (air-conditioned) (SEER/SSEER)	N/A	4.5/3.6	4.5/3.6
Cooling (mixed mode) (SSEER)	N/A	2.7	2.7
Heat recovery efficiency (%)	70%	70%	70%
Variable speed control of fans and pumps, controlled via multiple sensors	Yes	Yes	Yes
Demand control (mechanical ventilation only). Variable speed control of fans via CO_2 sensors	Yes	Yes	Yes

Note: FF = frame factor

- Industrial sites, workshops and non-residential agricultural buildings with low energy demand, i.e. where there is generally no heating or cooling or where any heating or cooling is for short periods of the year.
- Stand-alone buildings other than dwellings with a total useful floor area of less than 50 m².
- Some conservatories and porches.

Appendix D recommends the documentation that could be given to the Building Control Body in the form of reports to show evidence of compliance, e.g. specifications and products used in U-values entered in the BER calculations. Appendices E and F list the documents and standards referred to within the Approved Document.

Approved Document L2B: Conservation of fuel and power in existing buildings other than dwellings

Sections 1, 2 and 3: Introduction, the requirements and general guidance

As with the other Part L Approved Documents, the general Building Regulation guidance is covered in the opening sections. The types of building work covered by Approved Document L2B, on buildings other than dwellings, are:

- consequential improvements;
- certain extensions
- a material change of use
- a material alteration
- the provision or extension of a controlled service or fitting;
- the replacement or renovation of a thermal element.

Section 4: Guidance relating to building work

The construction of an extension triggers the requirement for a consequential improvement in buildings with a total useful floor area greater than 1000 m². In such cases, the guidance in Section 6 of L2B should be followed in addition to the following specific guidance.

Large extensions, with a total useful floor area greater than 100 m² and greater than 25% of the existing building, are regarded as a new building, and the guidance in Approved Document L2A should be followed. The requirement for a consequential improvement should also be met.

For **other extensions** the following performance standards should be achieved:

- Fit draught-proofed controlled fittings that meet the U-values in Table 13.10.
- Newly constructed thermal elements should meet the standards set out in Table 13.6.
- Existing opaque fabric that becomes part of the thermal envelope of the building whereas previously it was not should meet the standards in Table 13.7.
- The area of windows and rooflights in the extension should not exceed the values given in Table 13.11. An exception to this would be if the existing building has more openings, in which case it is reasonable not to exceed the existing proportions.

Where fixed building services are provided or extended as part of constructing the extension the guidance given below under work on controlled services and fittings should be followed.

An **optional approach** is to show that the area-weighted U-value of the elements in the extension are no worse than for a complying notional extension. SBEM calculations can be produced to show that the carbon dioxide emissions from the building including the proposed extension will be no worse than for the building with a complying notional extension. The building used in the calculation of both the notional and actual extension should incorporate the improvements proposed to meet the requirement for a consequential improvement. Where additional upgrades are proposed in the actual building to compensate for lower performance in the extension,

Table 13.10 Standards for controlled fittings

Fitting	Standard
Windows, roof windows and glazed rooflights	1.8 W/m² K
Alternative option in buildings of a domestic character	Window energy rating of C or 1.6 W/m² K
Plastic rooflight	1.8 W/m² K
Curtain walling	1.8 W/m² K or see alternative in note below
Pedestrian doors > 60% glazing	1.8 W/m² K
High-usage entrance doors for people	3.5 W/m² K
Vehicle access and similar large doors	1.5 W/m² K
Other doors	1.8 W/m² K
Roof ventilators	3.5 W/m² K

Curtain walling U-value should be no greater than the better of 1.8 W/m² K or the value given by:

$$ulimit = 0.8 + ((1.2 + (FOL \times 0.5)) \times GF$$

where FOL is the fraction of opening lights and GF is the glazed fraction.

Table 13.11 Opening areas in the extension

Building type	Windows and personnel doors as % of exposed wall area	Rooflights as % of roof area
Residential	30	20
Assembly, offices, shops	40	20
Industrial and storage	15	20
Vehicle access doors, display windows and similar glazing	As required	N/A
Smoke vents	N/A	As required

the upgrades should fully comply with the requirements of the Approved Document. The U-value design limits must be adhered to. Note that the guidance for **conservatories** is the same as described above for extensions to dwellings.

Material change of use and change of energy status

Where controlled services or fittings are being provided or extended they should follow the guidance as above for new buildings, with the following additional comments:

• For lighting schemes, a control factor is applied to the calculation of the average luminaire-lumens/circuit-watt which should encourage the installation of better controls.
• Where the work includes the provision of a thermal element, achieve the appropriate U-value from Table 13.6 and meet the requirement for continuity of insulation.
• Where more than 50% of the area of a thermal element or 25% of the total building envelope is being renovated assess the U-value against the threshold values in Table 13.7 and upgrade to the required U-value if necessary (subject to technical, functional and economic feasibility).
• Any thermal element that is being retained should be upgraded to the required U-value if the existing U-value is worse than the threshold value.
• Any existing window (including roof window or rooflight) or door that separates a conditioned space from an unconditioned space or the external environment and that has a U-value that is worse than 3.3 $W/m^2\,K$ should be replaced unless it is a **display window** or **high-usage entrance door**.

The option is available to use SBEM calculations to show that the carbon dioxide emissions from the building, as it will become, will be no worse than if the general guidance for change of use and change of energy status had been followed.

Material alteration

The guidance when carrying out a material alteration is the same as given above for a material change of use, except there is no requirement to upgrade retained thermal elements. There is a requirement to upgrade the U-value of an existing element that becomes a thermal element where previously it was not, if the existing construction will not meet the relevant threshold U-value given in Table 13.7.

Work on controlled services and fittings

The provision or extension of controlled services should ensure that the plant, controls and the insulation of pipes, ducts, vessels and lighting systems meet the recommendations of the *Non-domestic Heating, Cooling and Ventilation Compliance Guide*, NBS, 2013.

When replacing central plant the efficiency of the new plant should not be less than that of the controlled service being replaced. If the new service uses a different fuel, then the efficiency of the new appliance should be multiplied by the ratio of the carbon dioxide emission factor of the fuel used in the appliance being replaced to that of the fuel used in the new appliance before making this check.

In order to limit air leakage, ventilation ductwork should comply with the specifications given in *DW/144*.

The standards for fixed internal lighting are as for new buildings, but it should also be noted that when the area covered by the new lighting system is less than 100 m², the work should still comply with the standards, but need not be notified to Building Control (refer to Schedule 4 of the Building Regulations).

When installing building service systems to existing buildings, energy meters should be installed following the recommendations outlined above for new buildings.

The guidance for commissioning of services is as for new buildings in L2A.

Where windows, roof windows, rooflights or doors are to be provided, i.e. controlled fittings, they should be draught-proofed units whose area-weighted average performance is no worse than that given in Table 13.10.

Section 5: Guidance on thermal elements

Where newly constructed thermal elements are provided they should meet the U-values given in Table 13.6 and be shown to have continuity of insulation by adopting accredited construction details or providing evidence using guidance in IP 1/06 and BR 497.

Where a **thermal element is being renovated** it should achieve the standard set out in column 2 of Table 13.7; however, where the works apply to less than 50% of the individual element or 25% of the total building envelope no upgrade should be required. Any upgrade should be technically and functionally feasible and provide a simple payback of fewer than 15 years.

Where a **thermal element is retained**, for example:

- where an existing thermal element is part of a building subject to a material change of use;
- where an existing element is to become part of the thermal envelope and is to be upgraded;
- where an existing element is being upgraded as a consequential improvement

the element should be improved to the U-value in Table 13.7 if its existing U-value is worse than the given threshold value, subject to the test for technical, functional and economic criteria.

Section 6: Consequential improvements

Article 6 of the *Energy Performance of Buildings Directive* has been implemented through Regulation 28, which introduced the notion of consequential improvements. These apply to an existing building with a total useful floor area over 1000 m² [other than in Wales where the 1000 m² limit does not apply], where the proposed building work consists of, or includes:

- an extenstion;
- the initial provision of any fixed building services; or
- an increase to the installed capacity of any fixed building services.

Where Regulation 28 applies, **consequential improvements,** in addition to the proposed building work (the **principal works**), should be made to ensure that the building complies with Part L, to the extent that such improvements are technically, functionally and economically feasible, which is taken to mean that they achieve a simple payback not exceeding 15 years.

Where a **building is extended**, a way of complying with Regulation 28 would be to carry out **improvements that in ordinary circumstances are practical and economically feasible,** such as those listed below. These would only be to the extent that their value is not less than 10% of the value of the principal works, calculated by a suitably qualified person, i.e. chartered quantity surveyor, using current prices:

- Upgrading heating, cooling and/or air handling systems more than 15 years old by the provision of new plant or improved controls.
- Upgrading general lighting systems that have an average lamp efficacy of less than 40 lamp-lumens per circuit-watt and that serve areas greater than 100 m² by the provision of new luminaires or improved controls.
- Installing energy metering following the guidance given in CIBSE *TM39*.
- Upgrading thermal elements that have U-values worse than the threshold values in Table 13.7.
- Replacing existing windows, roof windows or rooflights (but excluding display windows) or doors (but excluding high-usage entrance doors) that have a U-value worse than 3.3 W/m² K.
- Increasing the on-site low and zero carbon (LZC) energy generating systems if the existing on-site systems provide less than 10% of on-site energy demand; the simple payback is reduced to seven years or less for this improvement owing to the typical higher cost and risk.
- Measures specified in the recommendations report produced with a valid Energy Performance Certificate.

Consequential improvements on installing building services can now be considered. On installing a fixed building service as a first installation, or where the installed capacity per unit area of an existing service is increased, it is necessary first to improve the parts of the building served by the service, where this is economically feasible, and second, make improvements in line with the guidance above for extensions. The cost of any improvement required to upgrade the areas of the building served by the service is not limited to 10% of the cost of the principal works and cannot be taken as contributing to the value of the second set of consequential improvements. The installed capacity of a fixed building service is defined as the design output of the distribution system output devices (the terminal units) serving the space in question, divided by the total useful floor area of that space.

In normal circumstances the following would be considered to be economically feasible. Where the installed capacity per unit area of a heating system is increased:

- upgrade thermal elements in the area to be served if the threshold U-value is worse than given in Table 13.7; and
- replace existing windows, roof windows or rooflights and doors (excluding display windows and high-usage entrance doors) within the area served that have U-values worse than 3.3 W/m² K with fittings that meet the U-values in Table 13.10 for replacement fittings.

Where the installed capacity per unit area of a cooling system is increased:

- upgrade thermal elements in the area to be served if the threshold U-value is worse than given in Table 13.7; and
- if the area of windows, roof windows (excluding display windows) within the area served exceeds 40% of the facade area or the area of rooflights exceeds 20% of the area of the roof and the design solar load exceeds 25 W/m^2, then the solar control provisions should be upgraded such that at least one of the following three criteria is met:

  (a) the design solar load is no greater than 25 W/m^2;
  (b) the design solar load is reduced by at least 20%;
  (c) the effective g-value is no worse than 0.3; and

- any lighting system within the area served by the relevant fixed building service that has an average lamp efficacy of less than 40 lamp-lumens per circuit-watt should be upgraded with new luminaires and/or controls.

Section 7: Providing information

A building log book should be provided as discussed for new buildings and should include details of new thermal elements, controlled services and fittings.

Chapter 14

Approved Document M: Access to and use of buildings

The title confirms that Approved Document M applies to all people, not just those with a disability. This shift of emphasis takes account of the Disabled Discrimination Act 1995 (DDA), the Disability Discrimination (Employment) Regulations 1996 (DDR), the wide range of 'disabilities' and the content of BS 8300. The DDA has been repealed and replaced by the Equality Act 2010, and supporting Regulations, but retains the obligations. The Act brings together more than 116 separate pieces of legislation forming a legal framework to protect the rights of individuals and advance equality of opportunity for all. Requirements M1–M4 can still be said to originate from the need to secure the welfare and convenience of persons in or about buildings. In Great Britain over six million adults have a disability of one form or another. They may have difficulty in walking, need to use a wheelchair or suffer from a hearing or sight impairment. This 'disability' may be temporary (e.g. a broken ankle) or long term (i.e. expected to last more than 12 months).

As before, compliance with the Workplace (Health, Safety and Welfare) Regulations 1992 and the Health and Safety at Work, etc. Act 1974 can be achieved by complying with Parts M and K.

The links between the guidance contained in Approved Document M and the duties of an employer or service provider are important. It is not a requirement to follow the guidance in the Approved Document (all editions) for satisfying the duties to make reasonable adjustments to physical features of the premises in certain circumstances.

All employers and *all service providers* must also take account of the duties under the Equality Act. Reasonable steps must be taken to remove, alter or provide a reasonable means of avoiding a physical feature of their premises that makes it unreasonably difficult or impossible for disabled people to make use of their services. Reference should also be made to Regulation 9, of the Equality Act 2010 (Disability) Regulations, in relation to a 10 year exemption.

For the purposes of Approved Document M specific definitions should first be studied:

Independent access – (from Regulation 2) to an extension or part of a building means access that does not pass through the rest of the building.

Access – approach, entry or exit.

Accessible – people, regardless of disability, age or gender, are able to gain access to buildings or parts of buildings.

Contrast visually – the visual perception of one element of the building, or fitting within the building, against another. Means that the difference in light reflectance value between the two surfaces is greater than 30 points. (Further information, *Colour, contrast and perception – Design guidance for internal built environments* – Reading University.)

Dwelling – a house or a flat (new blocks of flats built as student accommodation are to be treated as though they are hotel/motel accommodation in respect of space requirements and internal facilities).

Level – predominantly level, maximum gradient along the direction of travel of 1:60.

Principal entrance – the entrance that a visitor not familiar with the building would normally expect to approach.

Suitable – designed for use by people regardless of disability, age or gender, but subject to the usual gender-related conventions regarding sanitary accommodation.

Usable – convenient for independent use, with respect to buildings or parts of buildings.

The following meanings only apply to the sections on dwellings:

Common – serving more than one dwelling.

Habitable room – in relation to the principal storey, means a room used, or intended to be used, for dwelling purposes, including a kitchen but not a bathroom or a utility room.

Maisonette – a self-contained dwelling, but not a dwelling-house, which occupies more than one storey in a building.

Plot gradient – the gradient measured between the finished floor level and the point of access.

Point of access – the point at which a person visiting a dwelling would normally alight from a vehicle which may be within or outside the plot, prior to approaching the dwelling.

Principal entrance – the entrance which a visitor not familiar with the dwelling would normally expect to approach or the common entrance to a block of flats.

Steeply sloping plot – a plot gradient of more than 1:15.

Access strategy – may be provided at the time plans are submitted, and certainly if compliance with the Approved Document guidance cannot be followed. The access statement provided for planning purposes could act as a starting point, with issues added as the design/building work evolves. It is likely to be utilized a great deal for existing non-domestic building situations. Evidence, which may be sited within the statement, could include:

- adoption of BS 8300 guidance;
- use of recent validated research;
- agreement after consultations with Conservation or Local Access Officers, English Heritage or CADW;
- other convincing arguments that achieve an equivalent, or improved, solution;
- further guidance is to be made available from the Equality and Human Rights Commission (website www.equalityhumanrights. com).

BS 8300: 2009 (as amended) *Design of buildings and their approaches to meet the needs of disabled people – Code of Practice* provides in-depth guidance, beyond that contained in the Approved Document, for the design of domestic and non-domestic buildings, and based on user trials and validated desk studies. Slight variations exist between the two documents and therefore the Approved Document M should be taken as the starting point to ensure compliance with Requirements M1–M4. The Approved Document should be considered as the minimum standard and BS 8300 the best practice.

Requirement	Limits on application

PART M ACCESS TO AND USE OF BUILDINGS

Access and use

M1. Reasonable provision shall be made for people to –
(a) gain access to; and

(b) use the building and its facilities.

The requirements of this Part do not apply to –

(a) an extension of or material alteration to a dwelling; or

(b) any part of a building that is used solely to enable the building or any service or fitting in the building to be inspected, repaired or maintained.

Access to extensions to buildings other than dwellings
M2. Suitable independent access shall be provided to the extension where reasonably practicable.

Requirement M2 does not apply where suitable access to the extension is provided through the building that is extended.

Sanitary conveniences in extensions to buildings other than dwellings
M3. If sanitary conveniences are provided in any building that is to be extended, reasonable provision shall be made within the extension for sanitary conveniences.

Requirement M3 does not apply where there is reasonable provision for sanitary conveniences elsewhere in the building, such that people occupied in, or otherwise having occasion to enter, the extension can gain access to and use those sanitary conveniences.

Sanitary conveniences in dwellings
M4. (1) Reasonable provision shall be made in the entrance storey for sanitary conveniences, or where the entrance storey contains no habitable rooms, reasonable provision for sanitary conveniences shall be made in either the entrance storey or principal storey.
(2) In this paragraph 'entrance storey' means the storey which contains the principal entrance and 'principal storey' means the storey nearest to the entrance storey which contains a habitable room, or if there are two such storeys equally near, either such storey.

Before moving on to the technical guidance contained in the Approved Document, the first task must be to establish whether the requirements apply to the building work in question. The General Guidance Section 0 of the Approved Document makes it clear that the requirements of Part M apply to the majority of building work, namely:

- new non-domestic **building or dwelling;**
- an **extension** or **material alteration** of a non-domestic building;
- a **material change of use** to a hotel or boarding house, institution, public building or shop (including a restaurant, bar or public house).

The only areas where the requirements do not apply are listed under the *Limits of application*. This significant extension of the scope of Part M gives rise to additional guidance in Section 0. This can be summarized as follows:

- The extension of a non-domestic building should be treated as a new building, with suitable independent access where **reasonably practicable** or suitable access provided via the existing building.
- Where sanitary conveniences are provided in any building that is to be extended, reasonable provision must be made within the extension or within the existing building (where suitable access must be available).
- For a **material alteration,** the work itself must comply with Requirement M1. The work would include new, and alterations to, relevant features, e.g. entrances or access arrangements to other levels. **Reasonable provision** must be made for access and use of new or altered sanitary conveniences and access from the site boundary, or on-site car parking, must be no less compliant following a material alteration. (Note that this access route need not be upgraded to fully comply with M1, although reference must be made to the Equality Act.)
- In the case of a **material change of use** the building must be upgraded to comply with M1.
- For a part change of use **reasonable provisions** must be made to ensure compliance with M1 for that part, access to it and use of any sanitary conveniences provided (or to gain suitable access to them in another part of the building).
- To justify that the proposal offers **reasonable provisions** an **Access Strategy** will need to be provided.
- The note is made that Part M applies to features, outside the building, which are needed to allow access to the building from the edge of the site and from car parking and setting down points within the site.
- Although Part M does not apply to an extended or materially altered dwelling, compliance for the original dwelling must not be made more unsatisfactory than before. For example, a front porch extension may include the removal of the only WC at that storey level and worsen the entrance arrangements.

- For **educational establishments the guidance contained in the DfES 1997 Constructional Standards** (as described in Circular DfES/0142/2001) is now contained in the Approved Document. Note that purpose-built student living accommodation, including that in the form of flats, should be treated as hotel/motel accommodation in respect of space requirements and internal facilities.
- **Historic buildings** include listed buildings; those within a conservation area; buildings of architectural and historical interest (referred to as a material consideration in a Local Authority's development plan) including those within national parks, areas of outstanding natural beauty and world heritage sites; and vernacular buildings of traditional form and construction.
- The Approved Document recognizes the need to conserve historical buildings and to improve accessibility where **practically possible**. The advice and consultation with the Local Authority's conservation and access officers, and English Heritage or CADW: Welsh Historic Monuments, will help to achieve this aim. Examples include restoration of original features; rebuilding a historic building after fire; use of traditional construction methods; and sympathetic inclusion of lifts to certain levels.

Once it has been established that the requirements will apply to the building, then reasonable provisions should be made to enable people, regardless of disability, age or gender, for buildings other than dwellings, to:

- reach the principal entrance to the building (and other entrances described in the Approved Document) from the site boundary, on-site car parking and other buildings on the same site, e.g. university campus, school or hospital site;
- not be at risk of hazards caused by building elements, especially people with impaired sight, but rather be assisted by them in wayfinding;
- gain access into, and within, any storey of the building and to the building's facilities, subject to the usual gender-related conventions regarding sanitary accommodation;
- use suitable audience or spectator seating;
- use aids to communication for people with an impairment of hearing or sight (in auditoria, meeting rooms, reception areas, ticket offices and at information points); and
- use sanitary accommodation provided for the users of the building;

and, for dwellings, to:

- reach the principal, or suitable alternative, entrance to the dwelling from the point of access;

- gain access into and within the principal storey of the dwelling; and
- use the WC provision, at no higher storey than the principal storey.

Sections 1–10 tackle each of these performance criteria in a logical sequence. Part M does not extend to means of escape, where reference must be made to Part B. For example, an escape stair or route used for escape only need not comply with the Part M guidance. In mixed-use developments the requirements for non-domestic use should apply in any shared parts of the building.

Section I: Access to buildings other than dwellings

The provisions of this section recognize the differing needs of all people, including wheelchair users, ambulant disabled people and those with impaired sight, and are illustrated in Figure 14.1. The detailed design considerations are summarized for each element forming part of the route up to the building:

- Where possible a level approach should be provided, from the site boundary and designated car parking to the entrance(s), which should be at least 1500 mm wide. Passing places should be provided at maximum 50 m centres (or within sight of each other) 1800 mm wide × 2000 mm long and no obstructions to a height of 2100 mm.
- Approach gradient no steeper than 1:60 or 1:20 where landings provided for each 500 mm rise, maximum cross-fall gradient of 1:40.

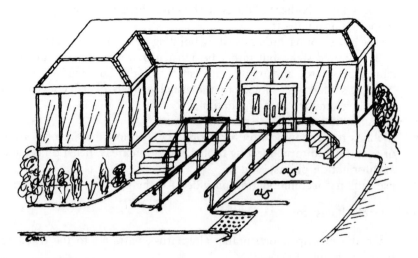

Figure 14.1 Principal provisions for approach up to building.

- Route to be firm, durable, slip resistant, undulations maximum 3 mm and joint gaps maximum 5 mm. Differing surface materials to have similar frictional characteristics; avoid use of gravel or sand.
- Entrance(s) to be lit and clearly identified.
- Pedestrian routes to avoid vehicular routes and incorporate dropped kerbs at uncontrolled crossings, identified by buff colour blister surface.

On-site parking is now specifically controlled under Building Regulations, although Town Planning conditions, BS 8300 and the Equality Act may influence the degree of provision:

- Provide minimum one designated disabled parking bay, on firm and level ground (as route above), as close as feasible to the principal entrance of the building. Overall dimensions 6000 mm × 3600 mm wide, incorporating a 1200 mm access zone to side and rear of the car, with dropped kerb access to the pedestrian route.
- Ticket machines, where necessary, to be accessible for wheelchair users and people of short stature, controls between 750 and 1200 mm above the ground.
- A setting down point is also required, clearly sign-posted on firm and level ground as close as practicable to the entrance(s).

For an approach steeper than 1:20 a **ramped approach** should be adopted. This needs to incorporate suitable complementary steps where the rise of the ramp exceeds 300 mm (2 × 150 mm steps), based on the following guidance:

- It should be readily apparent or clearly sign-posted.
- It should not be steeper than 1:20 (flight 10 m, maximum rise 500 mm), 1:15 (flight 5 m, maximum rise 333 mm) or 1:12 (flight 2 m, maximum rise 166 mm). No flight should have a going greater than 10 m, or a rise of more than 500 mm.
- Provide a lift where total rise is greater than 2 m.
- It should be minimum 1500 mm wide.
- It should provide a slip-resistant surface, contrasting colour to landings, but with similar frictional characteristics.
- Provide landings at the foot and head of the ramp, with an unobstructed length of at least 1200 mm. Intermediate landings to have an unobstructed length of at least 1500 mm (1800 mm square if a wheelchair user is unable to see from one end of the ramp to the other or the ramp has three flights or more). All landings should be level or a maximum gradient of 1:60, maximum cross-fall gradient of 1:40.
- Provide suitable handrail on both sides.
- Provide a minimum 100 mm kerb, with visual contrast, to open-sided flights and landings (in addition to any Part K guarding required).

The provision of a **stepped approach** is in addition to any ramped approach and should be based on the following guidance:

- Provide a level and unobstructed landing (minimum 1200 mm long) with 'corduroy' hazard warning surface at top and bottom of each flight.
- The steps should be a minimum 1200 mm surface width.
- No rise between flights to exceed 12 risers (for going less than 350 mm) or 18 risers (for going more than 350 mm). Single steps are not acceptable.
- Distinguish all nosings (on tread and riser) with permanent contrasting material, minimum 55 mm wide. Maximum step nosing projection should be 25 mm.
- The rise of steps should be between 150 and 170 mm (subject to agreement with building control body for existing situations) and the goings between 280 and 425 mm. Open risers to be avoided.
- Provide suitable continuous handrails on both sides of flights and landings, and divide into 1 m channels for widths over 1.8 m.

Handrails for both ramped and stepped approaches (32–50 mm diameter, 50–75 mm from wall) should be grippable, slip resistant and not cold to the touch and contrast visually against background; securely fixed at a height of between 900 and 1000 mm above the ramp or pitch line of the steps, 1100 mm to landings, and continuous across flights and landings. A second lower handrail, at 600 mm, should be considered for relevant buildings, e.g. schools. They should terminate to prevent clothes catching and extend 300 mm beyond the top and bottom of the ramp or step nosings, and not project into a route of travel.

Section 2: Access into buildings other than dwellings

The defined point(s) of access into the building for visitors, customers and staff (which may be specifically for their use) should all be suitable and accessible. For an existing building and where sloping ground or space restrictions present problems at the principal entrance a suitable alternative entrance may be provided. This would be subject to sufficient internal circulation existing in the building to allow access to the principal entrance. This principle also applies where access from disabled car parking cannot be achieved up to the principal entrance. The risks to people on entering the building should also be reduced. The provisions can be detailed as follows.

Accessible entrances should follow these guidelines:

- Clearly sign-posted (with the International Symbol of Access) and identified (against other building elements) from the edge of the site and non-accessible principal entrances.
- Have a level landing, clear of any structural supports and door swings, minimum 1500 × 1500 mm of a material that does not impede the movement of wheelchairs.
- Entrances with manually operated doors should be provided with weather protection.
- Level threshold or a maximum of 15 mm total height.
- Door entry systems to be accessible to deaf, hard of hearing people, and people who cannot speak.
- Internal floor surfaces and mat wells should be flush and not impede movement of wheelchairs or present a potential trip hazard.

Doors at accessible entrances should be accessible by all. For self-closing doors power operation must be considered, especially where opening force required at the leading edge would be greater than 30 N (0°–30°) 22.5 N (30°–60°). Minimum effective clear leaf width of 800 mm (750 mm for existing buildings) should be adopted, up to 1000 mm (775 mm) for buildings used by the general public. Specific design guidance for powered/non-powered doors and glazed doors can be listed as follows:

- Manually operated (non-powered) entrance doors to have 300 mm clear leading edge dimension.
- All door furniture, and controls, to contrast visually with door surface, operated by closed-fist, e.g. a lever handle, and not to be cold to the touch.
- Powered entrance doors should have slide, swing (open towards user) or folding action, controlled automatically or by push pad, swipe card, etc. (at 750–1000 mm control height).
- Suitable automatic sensors, dwell times, safety stops, visual/audible warnings and fail-safe operation to be incorporated where necessary.
- Glass entrance doors/screens should have suitable visually contrasting manifestation, reference Section 7 of Approved Document K.
- Avoid projection of a held open door opening onto an access route.

Entrance lobbies perform a range of practical functions but can present a hazard to various groups entering a building. Persons in a wheelchair or those pushing a buggy need to negotiate both sets of doors. With this in mind minimum clear lobby dimensions are given in the Approved Document.

- For single swing doors the minimum depth is basically 1570 mm + door leaf dimension/projection into the lobby.
- For double swing doors adopt 1570 mm + both door leaf dimensions/ projections.
- Minimum widths of 1200 mm, 1800 mm for double leaf doors, apply.

General provisions to adopt include avoiding distracting reflections; use of suitable floor/mat materials that should remove rainwater; and full-height projections more than 100 mm should have visually contrasting guarding.

Section 3: Horizontal and vertical circulation in buildings other than dwellings

Once access has been gained into the building consideration needs to be given to the convenient vertical and horizontal travel within the building for all. The main guidance provisions take account of the space requirements for manoeuvring a wheelchair and design features that make it possible for people to travel independently within buildings. The first point of contact and design feature to consider is the **entrance hall and reception area.**

- The reception point, where provided, should be identifiable and in sight from the entrance, have clear wheelchair approach to it and be sufficiently spaced away from the entrance to avoid external noise problems if applicable.
- A manoeuvring space of 1200 × 1800 mm, with 500 mm knee recess, should be allowed at the reception counter, or 1400 × 2200 mm if no knee recess. The counter itself should be 1500 mm wide, 760 mm high and a knee recess 700 mm high.
- All reception points should be provided with a hearing enhancement system. The floor surface in the locality should be slip resistant.

For certain purpose groups of buildings, e.g. a hotel, an alternative may be to have staff greet visitors, including those in a wheelchair, at dedicated and suitable locations in the reception area. The next key consideration is **internal doors** that should be avoided where possible, especially those on self-closing devices that need to be opened manually. This does, however, give rise to conflicts with fire safety and security issues. In general, the guidance above, relating to **doors at accessible entrances**, should be used for internal doors. This includes use of the same minimum widths and maximum opening forces.

In addition, fire doors, particularly those in corridors, should be provided with electro-magnetic hold-open devices, and those to individual rooms with swing-free self-closing devices. Note that the 300 mm leading edge dimension is not necessary for standard hotel bedrooms as well as power-controlled doors.

Corridors and passageways present the next hurdle for all persons using the building. Design guidance follows on from that provided for persons wanting to gain access to the building. They should be wide enough to allow wheelchair users, and other persons with disabilities, to pass all users of the circulation route.

- Projections, e.g. radiators, hose-reels or columns, must be avoided or provided with visually contrasting handrails.
- Where possible the unobstructed width should be 1800 mm, 1200 mm minimum. For widths less than 1800 mm, passing places, e.g. corridor junctions, should be provided at reasonable intervals, minimum 1800 mm wide.
- Floors should be slip resistant and level, or maximum 1:60 gradient, or designed as an internal ramp (as previously stated guidance for external ramps) if steeper than 1:20.
- Doors opening into major access routes must be avoided, unless they serve minor utility type rooms or a wheelchair-accessible toilet (where 1800 mm clear corridor width should be retained).
- Unequal door leaves on a major access route should be on the same side, glazed screens defined with manifestation, as previously stated in earlier guidance, consideration given to suitable acoustic design and good natural and artificial lighting provided.

For **internal lobbies** the general guidance for external lobbies should be followed.

Vertical circulation within the building

The most straightforward method of vertical travel between floor levels is by way of a **passenger lift**, and for certain building purpose groups this may be required as a matter of course, for example, a residential care home or a purpose-built office building. However, having in mind the cost and floorspace implications of providing a suitable lift, other options may need to be considered (see Figure 14.2), although it should be noted that no minimum net floor areas are now stated. In any case internal stairs, unless used for means of escape only, should be provided and designed for ambulant disabled and those with sight impairments.

PROVISION OF LIFTING DEVICES

- New developments should have a **passenger lift** serving all storeys.
- For new developments with 'site constraints' (e.g. listed building or infill site in a historic town centre) and existing buildings where a

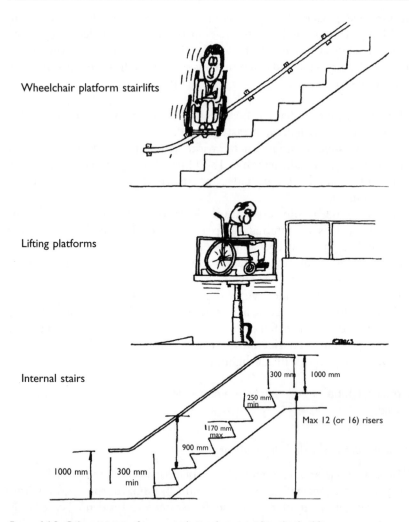

Wheelchair platform stairlifts

Lifting platforms

Internal stairs

Figure 14.2 Other options for vertical circulation within the building.

passenger lift cannot 'reasonably be accommodated' a **lifting platform** can be used.

- For other existing buildings, but only in exceptional circumstances, a **wheelchair platform stairlift** can be used to serve an intermediate level or a single storey.

General requirements for lifting devices can be summarized as follows:

- In front of each lifting device provide an unobstructed manoeuvring space of 1500 × 1500 mm, or 900 mm wide if straight approach.

- Call button height between 900 and 1100 mm, 500 mm or more from any return wall. Any symbols raised for tactile reading and buttons to contrast visually against face plate and face plate against wall.
- Floor not to be dark, have similar frictional qualities to landing and a handrail to be provided to at least one side, 900 mm nominal height and a suitable emergency communication system is to be installed.

Passenger lifts themselves should conform to the Lift Regulations 1997, e.g. by use of BS EN 81 – 70: 2003, and be accessible from the complete storey. Minimum lift car dimensions are 1100 × 1400 mm deep (mirror to be provided to allow wheelchair to reverse out of lift). Use of a larger car size, i.e. 2000 × 1400 mm wide, will allow a wheelchair to fully turn around and offer increased capacity. In addition:

- Provide a minimum 800 mm clear opening door width(s).
- Site car and landing controls between 900 and 1100 mm (preferred dimension), 400 mm or more from return wall.
- Car and landing doors to be visually distinguishable from adjoining walls and glass areas should be identifiable.
- Timing devices and reopening activators required to offer sufficient time for persons, including those with guide dogs, to enter and leave the lift car.
- A visual/voice indication of floor level reached and car arrival should be provided to car and landings.
- Evacuation lifts should accord with BS 5588: Part 8 (BS 9999, Section 5).

Lifting platforms represent the next method to transport wheelchair users, and their companions, between levels and storeys. They should comply with the requirements of the Supply of Machinery (Safety) Regulations 1992, i.e. by use of BS EN 81 series. Note that for travel height greater than 3 m a product certificate issued by a Notified Body would be required. Design should follow these provisions:

- Maximum vertical travel distance of 2 m if not enclosed and not penetrating a floor. Minimum platform dimensions 800 × 1250 mm deep, wheelchair user unaccompanied, 800 mm clear opening door width.
- Maximum vertical travel distance may exceed 2 m if enclosure is provided. Minimum platform dimensions 900 × 1400 mm deep, wheelchair user unaccompanied, 800 mm clear opening door width (1100 × 1400 mm deep where accompanied or where two doors at 90° to each other are provided, 900 mm clear opening door width required).
- Maximum rated speed of 0.15 m/s and full instructions for use to be provided.

- Site car continuous pressure controls between 800 and 1100 mm high, 400 mm or more from return wall.
- Doors to be visually distinguishable from adjoining walls and glass areas should be identifiable. They should also be accessible from the complete storey.
- A visual/voice indication of floor level reached and platform arrival should be provided to car and landings.

Wheelchair platform stairlifts can be taken as the last option for transportation of wheelchair users only, noting that they travel slowly which may not be appropriate for all users. It is used for conversions and alterations where it is not practicable to install a conventional passenger lift or a lifting platform. It may be suitable to 'small areas with a unique function', e.g. a small library gallery, a staff rest room or a training room. They should comply with the requirements of the Supply of Machinery (Safety) Regulations 1992, i.e. by use of BS EN 81 series. The specification and design of such stairlifts should follow these provisions:

- Unlike a domestic-type stairlift the minimum clear dimensions of the platform need to be 800 × 1250 mm deep, 800 mm clear access width. Note that the stairlift in the parked position must not restrict escape width in a single-stair building.
- Maximum rated speed of 0.15 m/s and full instructions for use to be provided, together with management supervision.
- Provide continuous pressure controls designed to prevent unauthorized use.

Internal stairs can in general follow the design guidance for stepped access as outlined earlier and by reference to Approved Document K, Section 1. Hazard warnings are not required.

Internal ramps can in general follow the design guidance for ramped access, excluding the provisions applicable to external locations and by reference to Approved Document K, Section 2.

Handrails to internal steps, stairs and ramps should follow the guidance in Approved Document K, Sections 1–3.

Section 4: Facilities in buildings other than dwellings

All facilities incorporated into the building design should be both accessible and usable for all people. The guidance in the amended 2004 edition of Approved Document M still does not address all possible facilities and it may therefore be necessary to refer to other guidance sources, e.g. BS 8300, to show compliance with the functional Requirement M1. The key issue here is that *all* people, including those with a disability or in a wheelchair, should be able to use *all* of the building facilities. Attendance as spectators, participants and staff should be made available at lectures, conferences, social venues, etc. This represents a significant extension to the scope of Part M and is explained within the following notes, which reflect the layout of the Approved Document guidance.

Audience and spectator facilities relate basically to seating provisions where segregated areas should be avoided. For **audience seating** wheelchair-accessible routes must be provided to wheelchair spaces, some in pairs.

- Minimum number of permanent spaces – 1% of capacity up to 10 000. Removable seating to provide remainder up to 6 for capacities of 600; above 600 but below 10 000 provide additional provision if desired.
- For capacities exceeding 10 000 refer to *Accessible stadia: a good practice guide to the design of facilities to meet the needs of disabled spectators and other users.*
- The horizontal space allocated for a wheelchair should be a minimum 900 × 1400 mm deep, minimum 900 mm clear access space/aisle dimension.
- Spaces to allow alternative views (especially where more than two spaces are provided), clear view of event and sign interpreters, adjoining standard seats to have lift-up arms and some to allow for assistance dogs to rest under or in front of seat.
- Stepped access routes should have handrails as previous guidance for steps. Note that the Approved Document offers guidance as to the location of wheelchair spaces in stepped terraced floors, namely at the accessible levels, i.e. top and bottom, only.
- For **lecture/conference** facilities a podium, including its equipment, should have full wheelchair access (ramp or lifting platform) and a suitable hearing enhancement system. This latter item should be applied to all such buildings; refer to later guidance notes.

Refreshment facilities – The significant change here is the objective that *all* users have access to all parts of the facility.

- Restaurants and bars, including changes in level, should be designed such that they can be reached and used by all people independently or with companions. Staff areas, all public areas, including lavatory accommodation, public telephones, external terraces (with suitable threshold), self-service and waiter service areas, should all be accessible.
- Bar or counter working surface or shared worktop facility (for example, a tea-point) should not be more than 850 mm above the floor, with

allowance for people standing and wheelchair approach; water delivery as later guidance.

The provisions for **sleeping accommodation** apply to hotel and motel bedrooms and also to student accommodation. The somewhat onerous requirement to provide 1 wheelchair-accessible bedroom in 20 (or part thereof) remains. The bedrooms must be located on accessible routes, linked to the remainder of the building and should have a standard level of amenity similar to other bedrooms. A choice of location should be offered, although for certain types of accommodation it may not be necessary on upper floors where they are no less advantageously situated. In student accommodation consideration needs to be given to providing a wheelchair-accessible toilet for visitors.

The following provisions apply to *all* bedrooms:

- Entrance and en suite doors to accord with previous guidance for **internal doors**.
- Internal wardrobe doors, etc. should open through 180°, with contrasting and easy grip handles.
- Openable windows and easy-to-use controls located between 800 and 1000 mm from floor.
- All bedrooms to have visual fire alarm signal, and embossed number on entrance door.

For specific wheelchair-accessible bedrooms the following additional items must be considered:

- Diagram 17 of the Approved Document gives just one example of a wheelchair-accessible bedroom; see Section 5 for details of wheelchair-accessible bathroom or shower. A 1500 × 1500 mm wheelchair turning space and 700 mm gap between beds should be provided to allow space to manoeuvre and transfer independently to the bed.
- Entrance door should incorporate wide-angle viewer, 1050–1500 mm.
- An emergency assistance alarm, activated by a red pull cord with a reset button, should be provided at the bed and adjacent floor area, with call signal outside the bedroom.
- Any balcony provided should be accessible, with suitable threshold and permanently clear zone of 1500 mm back from access doors.

The use of **switches, outlets and controls** has now been extended to *all* buildings. The design guidance basically extends that for dwellings and takes account of various workplaces and public use. The main provisions can be listed as follows:

- In general, wall-mounted socket outlets, telephone points and TV sockets are to be located between 400 and 1000 mm above floor level, permanently wired appliances between 400 and 1200 mm, between 750 and 1200 mm where precise hand movements are necessary and not more than 1200 mm where push button controls for limited dexterity are provided.
- Controls requiring close vision, e.g. thermostat, to be located between 1200 and 1400 mm above floor level.
- Locate socket outlets consistently in relation to doorways and room corners, light switches should align horizontally with door handles, 900–1100 mm, and have large push pads where used by the general public (or use pull cords), be easy to operate, indicate 'on' position (on and off for a circuit isolater) and have front plates contrast visually with their backgrounds.

Aids to communication are particularly important to enable persons with impaired hearing to attend public performances or to fully take part in a discussion. These should be suitably planned and provided in *all* rooms/ spaces used for meetings, lectures, classes, performances, spectator sport or films, and at service or reception counters in noisy areas or separated by glazed screens. The three main systems, which should be permanent, in common use are:

- **Loop induction** – where the amplified signal is directed through a loop around the room/space and picked up by a listener's hearing aid, reference BS 6083. For small rooms a portable system may be used.
- **Infrared** – which requires a personal receiver to demodulate and convert the signal in the invisible light.
- **Radio** – similar to infrared but using radio waves. **Sound field** systems are also now starting to be used, especially in schools.

The presence of such a system should be identified by the standard symbol, and suitable telephones should be marked with the 'T' symbol. Artificial lighting should also be compatible with such systems.

The only remaining provision within the Approved Document is the need for a clearly audible public address system, supplemented by visual information. Consideration should also be given to:

- appropriate choice of floor, wall and ceiling materials, avoiding shiny surfaces and large repeated patterns;
- public address system to take account of acoustic environment;
- careful design of artificial lighting to avoid glare, strong shadows and low-level up-lighting, but give good colour rendering.

These latter items are not listed as specific provisions, although for detailed design guidance BS 8300 is given as a reference source.

Section 5: Sanitary accommodation in buildings other than dwellings

The availability of sanitary conveniences should in principle be the same for all persons, wheelchair users, ambulant disabled, people of either sex with babies and small children, and even people encumbered by luggage. Consistent locations will also help people with learning difficulties, and the issues faced by those with visual or hearing impairments and tactile sensitivity problems must be considered.

General sanitary accommodation provisions, for WC compartments, wheelchair-accessible toilets and changing/shower rooms, are first highlighted.

- Taps for baths and wash hand basins should have lever-type (closed-fist) or automatic operation and terminal fittings should comply with Guidance Note G18.5 of the Guidance Document relating to Schedule 2: Requirements for water fittings, of the Water Supply (Water Fittings) Regulations 1999.
- Ironmongery to comply with guidance for **internal doors** and have an emergency release mechanism to allow door to open outwards.
- Opening doors not to obstruct escape routes, fire alarm to emit visual and audible signal and emergency assistance alarm system also to be provided.
- Exposed surfaces of radiators, etc. should be below 43 °C.
- Sanitary fittings, including grab rails, and wall/floor finishes themselves should contrast visually with background walls/floors.

The actual **provision of toilet accommodation** is the next important issue to consider.

- Where the building requires only one toilet, a small shop for example, then it should be unisex wheelchair-accessible with extra width (increased from 1500 to 2000 mm) to cater for a standing height basin.
- Wherever toilets are provided in a building, for visitors, customers and staff, at least one unisex wheelchair-accessible toilet should be provided.
- Separate-sex toilets should incorporate at least one ambulant disabled toilet.
- Separate-sex toilets, with four or more cubicles, should incorporate at least one enlarged cubicle.

We can now move on to the design guidance for the various types of toilet accommodation, the first being **wheelchair-accessible unisex toilets**. To satisfy Requirement M1 or M3 one should be located as close to the entrance/reception area as possible, not compromise privacy and be positioned on accessible routes, in a similar location on each floor of a multi-storey building (left/right transfer on alternate floors). The combined horizontal travel distance for any wheelchair user should not exceed 40 m (100 m if any doors on hold open devices) where unisex accommodation is provided on the same floor or via the use of a passenger lift (limited to one storey where a lifting platform only is provided).

A unisex wheelchair WC compartment should be at least 1500 × 2200 mm deep and contain the necessary equipment and fittings as shown in Diagrams 18, 19 and 20 of the Approved Document.

For **toilets in separate-sex washrooms** the guidance takes account of the needs of all people. Standard WC compartments should allow for a 450 mm diameter manoeuvring space behind an inward opening door and for ambulant compartments (minimum 800 mm width and 750 mm activity space) the guidance contained in Diagram 21 of the Approved Document should be followed, WC to accommodate a variable-height toilet seat riser. In addition, an enlarged compartment for those who need extra space should be provided (minimum width 1200 mm), with horizontal grab bar adjacent to WC, a vertical grab bar on rear wall and space for a shelf and fold-down changing table. Where wheelchair access is provided the WC compartment should have the same layout as a unisex facility and at least one washbasin (rim 720–740 mm above the floor) and at least one urinal (for men, rim set at 380 mm above the floor, two 600 mm vertical grab bars, centre lines at 1100 mm above the floor, either side of the urinal).

Wheelchair-accessible changing and shower facilities should be provided to all buildings, not just swimming pools and other recreational buildings, allowing disabled people to use the facilities independently or be assisted by others when necessary. The specific extent of the provisions is not stated, just the guidance where they are provided. Though for large shopping malls/retail parks and sports centres there should be one wheelchair-accessible unisex toilet provided with an adult changing table. Combined facilities need to take account of 'wet' and 'dry' zones and any individual accommodation, which is normally preferred, containing a WC, should not be the only WC available. The provisions to follow are grouped under four headings.

For changing and shower facilities

- Provide left- and right-hand transfer layouts where more than one individual compartment is available, with drop-down support rails and slip-resistant tip-up seat. Include emergency assistance systems and storage for amputee limbs.

- Communal shower facilities and changing facilities should have same space/equipment configuration as self-contained facilities but without doors.
- Sports facilities should have individual self-contained shower facilities and changing facilities in addition to communal separate-sex facilities (for sports buildings refer to Accessible sports facilities 2010 (Sport England)).

For changing facilities

- Individual self-contained changing facilities, including those for clothes shops, should comply with Diagram 22 of the Approved Document, minimum 2200 × 2000 mm wide (1500 mm manoeuvring space), with suitable grab rails and tip-up seat.

For shower facilities

- Individual self-contained shower facilities, and those provided in commercial premises for staff (at least one to be provided), to comply with Diagram 23 of the Approved Document, minimum 2200 × 2000 mm wide (1500 mm manoeuvring space), with suitable drop-down rails, shelf, curtain and tip-up seat.

For shower facilities incorporating a WC

- Provide left- and right-hand transfer layouts where more than one individual compartment is available, with fittings, etc. as outlined in Diagram 24 of the Approved Document, minimum 2500 × 2400 mm wide.

Wheelchair-accessible bathrooms are required for a range of buildings, including hotels, motels, student accommodation, overnight accommodation provided in hospitals, and bathing facilities in sports centres/health clubs. Layouts should allow for independent bathing or with assistance. Refer to earlier guidance in relation to en suite bathrooms to hotel bedrooms. The provisions follow previous principles:

- Provide left- and right-hand transfer layouts where more than one bathroom, with corner WC, is available, with fittings, etc. as outlined in Diagrams 25 and 26 of the Approved Document. Minimum dimensions 2500 × 2700 mm wide, incorporating 1500 mm manoeuvring space, 400 mm deep transfer seat and emergency assistance system.

Section 6: Means of access to and into the dwelling

Reasonable provision should be made within the boundary of the plot for disabled persons to approach and gain access into a dwelling from the point of alighting from a vehicle, which may be outside the plot. Note that the location and layout of dwellings/plots on a site will be a Town Planning matter.

This may be achieved by:

Approach to the dwelling

- A **level approach** with a gradient not exceeding 1:20 with a firm even surface, not loose (e.g. gravel), not less than 900 mm wide.
- A **ramped approach** that:

  (a) has a surface that is firm and even;
  (b) has flights of unobstructed width not less than 900 mm;
  (c) has flights not longer than 10 m for gradients up to 1:15, or 5 m for gradients not steeper than 1:12;
  (d) has top and bottom landings, and if necessary intermediate landings, with a length clear of any door or gate swings not less than 1.2 m.

  Note – handrails and kerbs are not required, but see Approved Document K where forming part of a building.

- A **stepped approach** where the point of access to the entrance has a gradient exceeding 1:15, which:

  (a) has flights with unobstructed widths not less than 900 mm;
  (b) has a rise of any flight between landings of not more than 1.8 m;
  (c) has top and bottom and if necessary intermediate landings not less than 900 mm;
  (d) has tread nosings as described for buildings other than dwellings, with the rise of each step being uniform and between 75 and 150 mm;
  (e) has steps with goings not less than 280 mm, this is measured 270 mm from the 'inside' for tapered treads; and
  (f) has a suitable continuous grippable handrail (where three or more risers) on one side of the flight 850–1000 mm above the pitch line and extending 300 mm beyond the top and bottom nosings.

- A **driveway** that provides an approach past any parked cars meeting the criteria of the above three methods of approach.

Access into the dwelling

Where the approach to the dwelling or block of flats is level or ramped an accessible threshold should be provided. An accessible threshold (showing compliance with Requirements C2 and C4) should also be provided to entrance level flats.

Where a stepped approach is unavoidable, an accessible threshold should be provided, and where a step into the dwelling is unavoidable the rise should not be greater than 150 mm.

An entrance door providing access for disabled people should have a minimum clear opening of 775 mm.

Section 7: Circulation within the entrance storey of the dwelling

Access should be facilitated within the entrance storey or the principal entrance storey into habitable rooms and a room containing a WC, which may be a bathroom on that level.

This is achieved through **corridors, passageways and internal doors within the entrance storey,** by way of doorways and corridor/passageways of minimum widths:

Doorway width	Corridor/passageway width
750 mm or wider	900 mm (when approached head-on)
750 mm	1200 mm (when approach not head-on)
775 mm	1050 mm (when approach not head-on)
800 mm	900 mm (when approach not head-on)

Local permanent obstructions, such as a radiator, are acceptable for a length not greater than 2 m providing the unobstructed width is not less than 750 mm for that length, and the obstruction is not placed opposite a door into a room so as to prevent a wheelchair user turning.

Where the plot is sloping a stepped change of level within the entrance storey may be unavoidable. **Vertical circulation within the entrance storey** may be achieved by reference to Approved Document K, Section 1.

Section 8: Accessible switches and socket outlets in the dwelling

In order to assist those with limited reach wall-mounted switches and socket outlets should be at suitable heights. That is outlets for lighting and other electrical appliances in habitable rooms should be between 450 and 1200 mm above finished floor level.

Section 9: Passenger lifts and common stairs in blocks of flats

Reasonable provision should be made to enable the disabled to visit occupants on any level. Ideally this should be by means of a lift, although this may not always be provided.

Where there is no passenger lift providing access between storeys a stair suitable for use by the disabled should be provided, and in any event a stair within a common area should be suitable for the visually impaired.

Where a lift is provided it should be suitable for use by an unaccompanied wheelchair user, and those with sensory impairments.

Provisions for lifts

The provisions for lifts to dwellings are essentially the same as for other buildings, as given in Section 3, with the exception of the lift car size. This should have a minimum load capacity of 400 kg and a width of at least 900 mm and length of at least 1250 mm (other dimensions may be used where test evidence, experience, etc. show that the lift will be suitable for unaccompanied wheelchair users).

Provisions for stairs

The provisions are now contained in Approved Document K, Section 1.

Section 10: WC provision in the entrance storey of the dwelling

The entrance storey of a dwelling should be provided with a WC, which may be within a bathroom, and which is located so that it can be reached from a habitable room without the need to negotiate a stair.

If there are no habitable rooms in the entrance storey a WC may be provided in either the entrance storey or principal storey.

Any WC compartment should:

- have a door that opens outwards and have a width as given in Section 7 above; and
- provide a clear space for wheelchair users to access the WC, and hand basin positioned so as not to impede access.

Dimensional criteria are given for **frontal access** and **oblique access**. In each case there should be a clear area from the front edge of the WC pan of at least 750 mm deep, and:

- for frontal access, 450 mm minimum (500 mm preferred) width either side of the centreline of the pan; or
- for oblique access, 450 mm minimum (500 mm preferred) measured from the centreline of the pan on the door side, and 400 mm on the opposite side; and

- also for oblique access, the edge of the door opening closest to the WC cistern should be positioned 250 mm behind the front edge of the pan (measured perpendicular to the centreline of the WC pan). For clarification, please refer to Diagrams 31 and 32 of the Approved Document.

In applying the guidance contained in Approved Document M, close cross-reference should be made to Approved Document B, concerning means of escape, and Approved Document K for the design and protection of stairs and ramps.

Chapter 15

Approved Document P: Electrical safety – dwellings

Published figures suggest that tighter regulation of electrical safety and competence of electrical contractors could be beneficial in the UK, although the number of fatalities is falling. The 2013 edition of the Approved Document takes account of the reduced range of electrical installations that are notifiable, but still aims to improve the design, installation, inspection, testing and certification of such installations. The Building Regulation Requirements only apply to new, extended or altered dwelling installations, avoiding overlap with existing electricity at work regulations, etc. Low-voltage and extra-low-voltage electrical installations should be protected against mechanical and thermal damage and avoid shock and fire hazard to people.

To ensure compliance with Requirement P1 an installer now has the choice of being registered with a **Competent Persons Scheme**, using a registered third party to certify the work or using a building control body.

PART P: ELECTRICAL SAFETY

Design and installation

Requirement
P1. Reasonable provision shall be made in the design and installation of electrical installations in order to protect persons operating, maintaining or altering the installations.

Limits on application
The requirements of this Part apply only to electrical installations that are intended to operate at low or extra-low voltage and are:
(a) in or attached to a dwelling;
(b) in the common parts of a building serving one or more dwellings, but excluding power supplies to lifts;
(c) in a building that receives its electricity from a source located within or shared with a dwelling; and
(d) in a garden or in or on land associated with a building where the electricity is from a source located within or shared with a dwelling.

Section I Design and installation

The provisions in this section can be summarized as follows:

- Professionally installed electrical work should comply with BS 7671: 2008 (as amended 2011). Information on operating, maintaining and altering the installation should be provided and include the following:
- certificate or report, with details/description of work carried out;
- permanent labels where required;
- operating instructions/logbooks; and
- detailed plans for large/complex installations.
- Note that Part P does not cover the functionality of electrically powered systems, i.e. fire alarms (Part B), fans and pumps (Parts F and L), etc.
- Cross-reference should be made to Part M for location of wall-mounted switches and socket outlets, consumer units provided with a child-proof lock/locked cabinet.
- Where a new dwelling is created by a material change of use, upgrading of the existing electrical installation may be necessary to meet current standards.
- Where alterations and/or additions are proposed, the new work should not adversely affect the existing installation or give rise to the unsafe operation of any new work. The existing installation should also be checked to ensure:
- rating/condition of the existing equipment to carry the additional loads;
- correct protective measures are used; and
- earthing and equipotential bonding arrangements are satisfactory.

Section 2: Application of Part P

The limits of application are relatively clear, although examples of application are given:

- Dwellings and flats, including those above business premises with a common supply.
- Common access corridors and staircases in blocks of flats, and shared amenities, e.g. laundries and gymnasiums.
- Parts of installations in or on land associated with the buildings (e.g. fixed garden lighting, air conditioning units, photovoltaic panels or pond pumps) and in outbuildings (e.g. sheds, garages and greenhouses).

From a building regulation point of view an electrical installation means fixed electrical cables or fixed electrical equipment located on the consumer's side of the electricity supply meter. A building control body should be notified of proposed electrical installation work as stated in Regulation 12(6A), namely:

- installation of a new circuit;
- replacement of a consumer unit; or
- addition/alteration of an existing circuit in a 'special location', i.e. within a bath/shower room, swimming pool or sauna.

All other work is *not* notifiable, although it should still be designed, installed, inspected, tested and certified to comply with BS 7671.

Section 3: Certification, inspection and testing

As stated in the opening paragraphs, one of three procedures should be chosen by the installer to certify that compliance is shown with Requirement P1.

Self-certification by a registered competent person

Installer to complete BS 7671 electrical installation certificate and provide copies to:

- the person ordering the work;
- the occupier (within 30 days of completion); and
- the building control body (within 30 days of completion).

Certification by a registered third party

Installer to appoint a registered third-party certifier to inspect/test/approve the work and issue an electrical installation condition report to:

- the person ordering the work (within five days of completion);
- the occupier (a Building Regulations compliance certificate within 30 days of completion); and
- the building control body (a copy of the certificate within 30 days of completion).

Certification by a building control body

Installer to notify a building control body, which will undertake the necessary inspections, or appoint a specialist to furnish it with a condition report. The competent installer will issue the BS 7671 certificate, and once approved, the building control body will issue a completion/final certificate to:

- the occupier (no time frame given).

The Approved Document concludes with Appendix A, key terms and Appendix B, the only British Standard referred to – namely BS 7671. The example circuit diagrams and BS 7671 forms are no longer in the Approved Document.

Further information

Addresses

Chartered Association of Building Engineers (ABE)
Lutyens House
Billing Brook Road
Weston Favell NN3 8NW
Tel: 0845 1261058
www.cbuilde.com

Association of Consultant Approved Inspectors (ACAI)
www.approvedinspectors.org.uk

Association for Specialist Fire Protection (ASFP)
Kingsley House
Ganders Business Park
Kingsley
Bordon
Hampshire GU35 9LU
Tel: 01420 471611
www.asfp.org.uk

British Board of Agrément (BBA)
Bucknalls Lane
Garston
Watford WD25 9BA
Tel: 01923 665300
www.bbacerts.co.uk

British Gas
PO BOX 4805
Worthing
BN11 9QW
www.britishgas.co.uk

British Standards Institute (BSI)
389 Chiswick High Road
London W4 4AL
Tel: 0208 996 9001
www.bsigroup.com

Building Research Establishment (BRE) [and the Fire Research Station (FRS)
and Loss Prevention Council (LPC)]
Garston
Watford WD25 9XX
Tel: 01923 664000
www.bre.co.uk

Building Services Research and Information Association (BSRIA)
Old Bracknell Lane West
Bracknell RG12 7AH
Tel: 01344 465600
www.bsria.co.uk

Centre for Accessible Environments (CAE)
Fourth Floor Holyer House
20-21 Red Lion Court
London EC4A 3EB
Tel: 020 7822 8232

Chartered Institute of Building (CIOB)
1 Arlington Square
Downshire Way
Bracknell
RG12 1WA
Tel: 01344 630 700

Chartered Institution of Building Services Engineers (CIBSE)
222 Balham High Road
London SW12 9BS
Tel: 0208 675 5211
www.cibse.org

Construction Industry Council (CIC)
26 Store Street
London WC1E 7BT
Tel: 0207 399 7400
www.cic.org.uk

Construction Industry Research & Information Association (CIRIA)
Griffin Court
15 Long Lane
London, EC1A 9PN, UK
Tel: 020 7549 3300 www.ciria.org

Department for Communities and Local Government (DCLG)
Eland House
Bressenden Place
London SW1E 5DU
Tel: 0303 444 0000
www.communities.gov.uk
See also Planning Portal at www.planningportal.gov.uk

Department for Education (DfE)
Castle View House
East Lane
Runcorn
Cheshire WA7 2GJ
Tel: 0370 000 2288
www.education.gov.uk [Moving to www.gov.uk]

Environment Agency
National Customer Contact Centre
PO Box 544
Rotherham S60 1BY
Tel: 08708 506506
www.environment-agency.gov.uk

Fire Protection Association (FPA)
London Road
Moreton-in-Marsh
Gloucestershire
GL56 0RH
Tel: 01608 812 500
www.thefpa.co.uk

Health and Safety Executive (HSE)
Tel: 0845 345 0055
www.hse.gov.uk

The Stationery Office
Tel: 0870 600 5522
www.tso.co.uk

The Institute of Clerks of Works and Construction Inspectorate of
GB Inc. (ICWCI)
28 Commerce Road
Lynch Wood
Peterborough
PE2 6LR
Tel: 01733 405160
www.icwgb.org

The Chartered Institute of Waste Management (CIWM)
9 Saxon Court
St Peter's Gardens
Marefair
Northampton NN1 1SX
Tel: 01604 620426
www.ciwm.co.uk

Institution of Civil Engineers (ICE)
1 Great George Street
London SW1P 3AA
Tel: 0207 222 7722
www.ice.org.uk

Institution of Fire Engineers (IFE)
London Road
Moreton-in-Marsh
Gloucestershire
GL56 0RH
Tel: 01608 812 580
www.ife.org.uk

Institution of Structural Engineers (ISE)
11 Upper Belgrave Street
London SW1X 8BH
Tel: 0207 235 4535
www.istructe.org

National House Building Council (NHBC)
Customer Services
NHBC
NHBC House
Davy Avenue
Knowlhill
Milton Keynes
MK5 8FP
Tel: 0800 035 6422
www.nhbc.co.uk

Oil Firing Technical Association (OFTEC)
Foxwood House, Dobbs Lane
Kesgrave 1P5 2QQ
Tel: 0845 6585080
www.oftec.org

Royal Institute of British Architects (RIBA)
66 Portland Place
London W1B 1AD
Tel: 0207 580 5533
www.architecture.com

Royal Institution of Chartered Surveyors (RICS)
12 Great George Street
London SW1P 3AD
Tel: 024 7686 8555
www.rics.org

Steel Construction Institute (SCI)
Silwood Park
Ascot SL5 7QN
Tel: 01344 636525
www.steel-sci.org

Timber Research and Development Association (TRADA)
Stocking Lane
Hughenden Valley
High Wycombe HP14 4ND
Tel: 01494 569600
www.trada.co.uk

United Kingdom Accreditation Service (UKAS)
21–47 High Street
Feltham
Middlesex TW13 4UN
Tel: 0208 917 8400
www.ukas.com

Index